Samuel Hubbard Scudder

Materials for a monograph of the North American Orthoptera

Samuel Hubbard Scudder

Materials for a monograph of the North American Orthoptera

ISBN/EAN: 9783337717728

Printed in Europe, USA, Canada, Australia, Japan

Cover: Foto ©ninafisch / pixelio.de

More available books at **www.hansebooks.com**

MATERIALS FOR A MONOGRAPH

OF THE

NORTH AMERICAN ORTHOPTERA.

BY

SAMUEL H. SCUDDER.

[From the Boston Journal of Natural History, Vol. VII. No. 3.]

CAMBRIDGE:
PRINTED BY H. O. HOUGHTON.
1862.

ART. VIII.— *Materials for a Monograph of the North American Orthoptera, including a Catalogue of the known New England Species.* By SAMUEL H. SCUDDER.

[Read May 21st, 1862.]

THE North American Orthoptera have been very much neglected, very little special attention having been paid to them; for besides the earlier general works of Linneus, Fabricius, De Geer, Stoll, and Palisot de Beauvais, and the more recent ones of Burmeister and Serville, which included descriptions of North American Orthoptera among others, the only other notices of our species have been in the few scattered descriptions by Say, Haldeman, Kirby, Girard, and myself, if we except only these two:— Harris's mention of the New England Orthoptera in his " Report on Insects of New England Injurious to

Vegetation," which, though very meagre indeed, is yet a fuller treatment of them than has been given by him to any other group in his Report; and De Saussure's short diagnoses of new species in the " Revue et Magasin de Zoologie," about three years since, which have reference principally to tropical forms.

My earliest intention in this paper was to restrict myself to a critical revision of the Orthoptera of New England, known to me, whether described or undescribed; and this has continued to be the main purpose of the article, in its present form, — more enlarged, because it was readily seen that a much better understanding might be obtained of the peculiarities of its fauna by comparisons, the more extended the better, with species closely allied from without its limits. This has been the case, particularly among the *Locustariæ;* and it is hoped that some better understanding may thereby be had of the Orthopteran fauna of the whole of North America.*

To further my purposes I solicited the assistance of many of my friends, and have invariably received their kindest coöperation. The original collection of Dr. T. W. Harris, in the Society's possession, has been invaluable to me; for by it I have been enabled to know exactly the extent of his knowledge of our fauna, as well as to determine his species directly from his types. The collection of the Museum of Comparative Zoölogy has been of great assistance to me, — containing as it does many species I could not otherwise have seen from the southern

---

* The whole number of species mentioned is 115, of which 78 are found in New England, distributed among the different families as follows: —

	Whole No.	From New England.
Forficulariæ	2	2
Blattariæ	9	7
Phasmida	1	1
Gryllides	14	11
Locustariæ	41	16
Acrydii	48	41
	115	78

and western portions of our country, — besides the New England collection of Mr. A. Agassiz. To Mr. P. R. Uhler, of Baltimore, — almost the only one in our country who studies the Orthoptera, — I am greatly indebted, both for many valuable suggestions, and for some very interesting insects from various parts of our Union, and especially for the opportunity of examining many species of *Ceuthophilus*, new to me. From Mr. F. G. Sanborn I have received very rich collections from Massachusetts, including the Orthoptera of the State Cabinet. My thanks are also due to Miss A. M. Edmands, of Cambridge; Messrs. C. A. Shurtleff, of Brookline, Mass.; Edward Norton, of Farmington, Conn.; and A. S. Packard, Jr., of Brunswick, Me., who have generously placed all their collections from their several vicinities in my hands, and to Mr. C. Thomas, of Murphysboro', Illinois, for many specimens from that State. By my own collections along the route taken by Sir John Richardson in Northwest America, I have been enabled to determine the few species described by Kirby, as well as to add others to its fauna, and by specimens collected for a number of years in the New England States, to add considerably to my material for this investigation.

Such have been my opportunities for the study of our Orthoptera, so far as native species are concerned; but there have also been of very material assistance to me in generic determinations, — the large series of European Orthoptera in the Cambridge Museum, — and a very fully represented and beautiful collection of European species which I have received from Herrn Brunner von Wattenwyl, who has in his possession the original collection which formed the basis of Fischer's elaborate work.

I have only made mention in this paper of species which I have myself seen, with but a single exception — *Ceuthophilus scabripes*, and have endeavored to verify every synonymical reference. Of the work of Stoll, I have only been able to make a hasty examination, without the opportunity of direct comparison with specimens; and the references to Palisot de Beauvais are not so complete as they would have been had I ever seen a perfect copy. In my references to Harris's Report, I have quoted the last (third) edition only, because it is better known, has illustrations of many species, and there has been no essential alteration in the text of the three editions. To those who would not otherwise understand the claim of prior description in some cases, I would say that the first edition was published in 1841. I have also made full references to his " Catalogue of the Insects of Massachusetts," (published in 1835, in Hitchcock's " Report on the Geology, Mineralogy, Botany, and Zoölogy of Massachusetts," and also in a smaller volume extracted from it, under the title, " Catalogue of the Animals and Plants of Massachusetts,") not because the names given there have any value in questions of priority, but because Dr. Harris had sent away many collections correspondingly labelled.

To prevent misunderstanding, I may state that where I have not indicated the number of specimens examined, it is to be taken for granted that I have seen a considerable number; and where no reference is made to the sexes seen, I have examined both. The measurements given have been, so far as possible, average measurements. The length of the body has been given as a straight line, connecting (in the natural attitude of the Orthopteron) the tip of the vertex and the extremity of the abdomen — excluding the ovipositor in the females of Gryllides and Locustariæ, and including the inarticulated abdominal appendages in the males; the length of the ovipositor is given as a straight line, connecting its tip with the point of junction of the upper and lower valves at its base.

For more easy reference I have prefixed an asterisk (*) to such species as I have seen from New England. In giving localities, I have only referred to those from which I have myself seen specimens, and have appended to them the names of the collectors; or where that was not known, the person from whom I received them.

I add here a Table for the more ready determination of the genera, which I have made as simple and concise as possible. It is strictly limited to the genera of New England, and is not intended for reference to the species outside its limits.

A TABLE TO FIND THE GENERA OF NEW ENGLAND ORTHOPTERA.

1. Hind legs longest; hind femora thickened 4
1. Legs of nearly equal length; hind femora not thickened . . . 2
2. Abdomen armed behind with a forceps . . . (*Forficulariæ*) 6
2. Abdomen not armed behind with a forceps 3
3. Body broad and flat . . . . (*Blattariæ*) 7
3. Body long and exceedingly slender, with slender legs. (*Phasmida*) Diapheromera.
4. Antennæ long and tapering 5
4. Antennæ short . . . . (*Acrydii*) 23
5. Wing covers with the outer portion bent abruptly downwards. (*Gryllides*) 11
5. Wing covers sloping downwards at the sides . . (*Locustariæ*) 15

FORFICULARIÆ.

6. Antennæ with less than 12 joints . . . . Labia.
6. Antennæ with more than 20 joints . . . . Labidura.

BLATTARIÆ.

7. Winged 8
7. Wingless . . . . Pycnoscelus.
8. Females with developed wings 9
8. Females with rudimentary wings . . . . . . Stylopyga.
9. Basal joints of tarsi longer than the others 10
9. Basal joints of tarsi shorter than the others . . . . . Ectobia.
10. Supraänal plate deeply fissured . . . . . Periplaneta.
10. Supraänal plate not deeply fissured . . . . Platamodes.

GRYLLIDES.

11. Fore-tibiæ broad 12
11. Fore-tibiæ slender 13
12. Insect large . . . . Gryllotalpa.
12. Insect small . . . . Tridactylus.
13. Hind-femora stoutish 14
13. Hind-femora slender . . . . Œcanthus.

14. Last joint of the maxillary palpi of the same length as the penultimate
                                                                    Gryllus.
14. Last joint of the maxillary palpi double the length of the penultimate
                                                                    Nemobius.

LOCUSTARIÆ.

15. Wingless, or with rudimentary wings and wing-covers 16
15. Winged 17
   16. Wingless; pronotum not extended over meso- and metanotum . Ceuthophilus.
   16. With rudimentary wings and wing-covers; pronotum extended
       over meso- and metanotum . . . . Thyreonotus.
   17. Wing-covers expanded in the middle 18
   17. Wing-covers not expanded in the middle 20
      18. Wing-covers much broadened in the middle, concave . . Cyrtophyllus.
      18. Wing-covers somewhat broadened in the middle, not concave . . 19
   19. Ovipositor very small . . . . Microcentrum.
   19. Ovipositor of medium size . . . . Phylloptera.
      20. Vertex of the head with a conical projection forwards . Conocephalus.
      20. Vertex of the head without a conical projection 21
   21. Ovipositor straight, or very nearly so; insect small . . Xiphidium.
   21. Ovipositor curved; insect large 22
      22. Ovipositor curved sharply upwards . . . . . Phaneroptera.
      22. Ovipositor ensiform . . . . Orchelimum.

ACRYDII.

23. Pronotum produced backwards, covering abdomen 31
23. Pronotum not produced backwards over abdomen 24
   24. Antennæ broad and flattened at base, acuminate . . . Opomala.
   24. Antennæ filiform, or slightly flattened 25
   25. Prosternum with a slender central spine 26
   25. Prosternum without a spine 28
      26. Sternal spine prominent 27
      26. Sternal spine but slightly raised . . . . Arcyptera.
27. Lateral carinæ of pronotum more or less prominent; medial carina scarcely
    elevated; extremity of abdomen of males much swollen . Caloptenus.
27. Lateral carinæ of pronotum wanting; medial carina generally prominent;
    extremity of abdomen in males not swollen . . . Acridium.
   28. Front, nearly perpendicular, generally swollen; vertex not prom-
      inent . . . . Œdipoda.
   28. Front considerably oblique, sloping inwards; vertex prominent . . 29
   29. Hind-border of pronotum sharply angulated . . . . Tragocephala.
   29. Hind-border of pronotum docked or rounded 30
      30. Foveolæ on the vertex; lateral carinæ of pronotum incurved . Stenobothrus.
      30. No foveolæ on the vertex; lateral carinæ of pronotum parallel or nearly
        so . . . . Chloëaltis.
   31. Pronotum arched roundly . . . . Batrachidea.
   31. Pronotum nearly or quite horizontal 32
      32. Antennæ 13-14 jointed . . . . Tettix.
      32. Antennæ 22 jointed . . . . Tettigidea.

FORFICULARIÆ, Latreille.

SPONGOPHORA, SERVILLE (emend.)

*1. S. BIPUNCTATA, nov. sp.

The head, antennæ, and prothorax are reddish brown; eyes black; elytra darker brown, with a rather large pale spot equi-distant from the base and either border; wings at rest, extending nearly twice as far back as the elytra, pale, with a dark brown band along the inner edge beyond the elytra; there is a faintly impressed longitudinal line on the prothorax. Length from front of prothorax to tip of wings, .3 in.

I place this species in this genus with some degree of doubt, because I have but a single mutilated specimen to examine, which wants abdomen and legs, the elytra and wings of one side, and the principal part of the antennæ; but the parts that remain exhibit good specific characters.

Mass. (H. Coll.) Taken May 30. 1 specimen.

LABIA, LEACH.

*1. L. MINUTA, nov. sp.

Thorax, elytra, and exposed portion of wings yellowish brown, covered with shortish hairs; middle of abdomen reddish brown; sides of the abdomen above and the head dark reddish brown approaching a black; last segment of abdomen and forceps reddish brown; abdomen also and forceps thickly beset with shortish hairs; legs shining pale yellow; parts of the mouth and antennæ yellowish brown; abdomen beneath brownish yellow; ♂ forceps slender, of nearly equal width throughout, curving outwards slightly at the middle, and then inwards towards the tip; slightly falciform, and meeting only at the tip; minute triangular black teeth on the lower inner edge; ♀ forceps diminishing in size to the tip, which is very slightly curved inwards; otherwise they are straight, meeting along the whole inner edge, which is toothed as in the ♂; tips as united together rounded, very slightly shorter than in the ♂. Length of whole body, .2 in.; forceps, .04 in. Mass. (H. Coll., November 4, 1845, flying in evening; Shurtleff, Aug. 28.) Virginia, (Uhler.) imperfect specimen; 3 ♂, 1 ♀.

BLATTARIÆ, Latreille.

Stylopyga, Fischer, having been proposed as a genus for the reception of *Blatta orientalis* L., before *Periplaneta* of Burmeister, must be retained for it; but since *B. americana* L. must be placed in a different genus from *B. orientalis*, we may reserve *Periplaneta* for that species with its allies, and the more properly since it is mentioned first under the genus by Burmeister. The two genera will be found to differ not only in the rudimentary condition of the wings in the females of *Stylopyga*, but also in the wings of the males, which are much shorter than the body in *Stylopyga*, and longer than the body in *Periplaneta*; the outer border also of the anterior pair is much less rounded in the former than in the latter, while another character is found in the deep fissuration of the supraänal plate in *Periplaneta*, which is almost wholly wanting in *Stylopyga*, where it is squarely docked, instead of being pointed, as in *Periplaneta*.

STYLOPYGA, FISCHER DE W.

*1. S. ORIENTALIS, Fisch. d. W., Bull. d. Nat. de Moscou; VI. 366. (1833.) For synonymy, see Fischer, Orth. Eur.

Seaboard of Mass., (H. Coll., Shurtleff, Sanborn, S. H. S.) N. York, (S. H. S.) Maryland, (Uhler.) The proportions of the length of the elytra to their breadth in this species and in *Periplaneta Americana* are very variable.

PERIPLANETA, BURMEISTER.

*1. P. AMERICANA, Burmeister, Handb. d. Ent.; II. 503. (1838.) For synonymy, see Fischer, Orth. Eur.; to which add ♀ *Blatta domingensis*, Pal. de Beauv., Ins.; 182, Pl. 1 b., fig. 4. (1805.)

Seaboard of Mass., (H. Coll.) Indiana, (Mus. Comp. Zoöl.) Mexico, (Uhler.) Texas, (Mus. Comp. Zoöl., Uhler.)

PLATAMODES, Nov. gen. (πλαταμώδης.)

A genus more closely allied to *Periplaneta* than to any other, but readily distinguishable from it by its much narrower and more elongated body, — the sides being sub-parallel to one another throughout their whole extent, while in *Periplaneta* the abdomen is much swollen. The wings and wing-covers extend beyond the abdomen, the latter being well rounded at the tip. The supraänal plate is regularly rounded, but lacks altogether the fissuration seen in *Periplaneta;* but at the same time it is not squarely docked as in *Stylopyga*. The anal cerci are somewhat shorter, and not so flattened as in *Periplaneta*, while the anal styles are very short and turned abruptly downwards. In *Periplaneta* the sub-genital plate does not extend so far backward as the supraänal. In *Platamodes* it extends backward farther. A further distinction between the two genera may be seen at the inner borders of the eyes, which in *Platamodes* are nearly parallel; while in *Periplaneta* they approach one another anteriorly. I have only seen males.

1. P. PENNSYLVANICA.

Blatta pennsylvanica, De Geer, Mem.; III. 537. Pl. 44, fig. 4. (1773.)
"   "   Oliv., Enc. Meth.; II. 317. (1791.)

Indiana, (Mus. Comp. Zoöl.) Maryland, (Uhler.) 5 ♂.

* 2. P. UNICOLOR, nov. sp.

Blatta pennsylvanica, H. Cat.; 56, (non De Geer.)

Blatta rufescens and *bicolor?* H. Cat.; (non Pal. de Beauv.)

Wings and wing-covers uniform pale shining reddish brown; head and prothoracic shield nearly the same, but slightly darker, particularly in the middle of the latter; abdomen a little darker above, especially on the borders; cerci dark brown; legs, especially the tibiæ, darker than the body; eyes black; antennæ and palpi brown; antennæ reaching backwards to tip of wing-covers. Length of body, .25 in.; length to tip of wings, .35.

"In woods, under stones, and entering houses by night in June." — H.

Mass., (H. Coll., Sanborn.) 6 ♂.

ECTOBIA, WESTWOOD.

*1. E. GERMANICA, Stephens, British Entomology; VI. 46. (1835.) For synonymy, see Fisch., Orth. Eur.; to which add *Blatta parallela*, Say Mss. and H. Cat.; 56.

Mass. (H. Coll., Shurtleff, Sanborn, S. H. S.) Vt., (Mus. Comp. Zoöl.) N. York, Maryland, (Uhler.)

*2. E. LITHOPHILA, Harris Mss.

I have not seen any winged individuals of this species. Immature specimens are of an uniform bright brownish red upon the head and thorax, with the abdomen growing darker posteriorly and laterally; the legs are of a bright shining yellow, a little dusky, with rather long delicate spines placed irregularly upon the edges, the femora being tipped with one slightly curved; cerci blackish; eyes black; palpi dusky; antennæ light brown; third joint of antennæ as long as the succeeding five taken together, and twice as large as the second; in *E. germanica* it is only slightly larger than either of the succeeding, and of the same size as the second. Length, .4 in.; breadth, .2 in. An öotheca of this species (or so labelled by Harris), is similar in general appearance to that of *E. germanica*, but is shorter; it measures — Length, .2 in.; breadth, .1 in.; thickness, .07 in. There are nine transverse depressions. "Under stones; very common in woods. April 20, 1834." — H.

Mass. (H. Coll., Shurtleff, Sanborn, S. H. S.)

*3. E. FLAVOCINCTA, nov. sp.

Prothoracic shield rather dark brown, slightly paler along the median line, bordered throughout with a pale yellowish band, forming only a very narrow edge posteriorly; broader in front, and quite broad at the sides, covering all the deflexed border; the edge at the sides and in front is slightly raised; wing-covers scarcely reaching tip of abdomen, reddish brown, with the anterior half of the outer margin paler, with a yellowish tinge; wings not half the length of the wing-covers; abdomen above very dark brown; below dark brown, the terminal segment being darkest; legs yellowish brown, with spines as in *E. lithophila*; head reddish brown; sides below antennæ yellowish; eyes black; antennæ dark brown, paler toward tip; third joint rather larger than the two succeeding joints, and equal in size to the second. Length of body, .56 in. "In woods, under a stone."— H.

Mass. (H. Coll.) Western States, (Mus. Comp. Zoöl.) Lake Superior, (Mus. Comp. Zoöl.) 3 specimens.

CRYPTOCERCUS, Nov. gen. (κρυπτὸς, κέρκος.)

A genus allied to, but very distinct from, *Polyzostæria*, Burm. The head is not large, much flattened, front sloping strongly inwards; prothorax broader than long, considerably arched, swollen, with the front border extending over the head slightly upturned to form a sort of hood; border of prothorax thickened in front considerably, at the sides more narrowly, here forming a slightly raised edge, which extends along the whole side of the body. Both sexes wingless; the meso- and metathorax not so arched as the prothorax; the hind border of the mesothorax not turned backwards at outer angle; that of the metathorax only in a very slight degree; abdomen not flattened, but quite full, especially on posterior half; the abdomen slightly longer than the thorax; the segments nearly equal in width, with the exception of the last (seventh) which is very large, triangular, three times the width of the sixth, produced posteriorly to a blunt rounded apex, the segment concealing the abdominal appendages altogether; the abdomen is a little longer than broad, regularly and but little rounded at the sides. The eyes are small, reniform, subglobose, the longitudinal diameter lying in the plane of the head; no ocelli; antennæ inserted in a broad circular depression, and about half the length of the body; first joint twice as long as the second, second as long as the third and fourth, third as long as the fourth and fifth; the terminal joints somewhat moniliform, the whole profusely covered with short hairs; third and fourth joints of the maxillary palpi equal, the last a little longer, considerably thickened at the termination. Legs compactly fitted to one another beneath the spreading sides of the thoracic segments; the femora broad, short, flattened, smooth, without a trace of spines, save one or two small ones at the tip upon the hind border, generally upon the upper edge only; tibiæ of fore legs very short and heavy, thickened at the tip; those of the other legs similar in character, but becoming longer posteriorly, — all thickly covered with heavy spines; tarsi with the first and last joint nearly equal in length, (on the anterior pair the last is much the longest) and equal to the three middle joints; well-developed claws, but with no pad between them; the abdominal appendages cannot be seen except through the gaping of the dorsal and ventral shields of the seventh segment; the cylindrical hairy cerci in both sexes are so long as just to reach the apex of the triangular supraänal plate; the styles of the ♂ are very small.

1. C. PUNCTULATUS, nov. sp.

Above of an uniform dark shining mahogany-brown color, a little deepest at the posterior extremity; beneath a little lighter, with a reddish yellow tint, especially upon the coxæ, and to be seen also on the mouth parts and the sockets of the antennæ; antennæ dirty brown; the whole body thickly covered with punctures, most abundant and largest upon the upper surface of the seventh segment, where also the edges are raised; beneath they are more distant; upon the head they are minute; there is a faintly impressed median line along the thoracic segments, excepting upon the hood-like projection of the prothorax; the anterior half of the raised edge of the sides of the prothorax is externally indistinctly grooved; the ♂ is a little more arched upon the thoracic segments than the ♀. The dorsal shield of the seventh segment is slightly excavated at the tip in both sexes; and the ventral shield in the ♀ is a little indented upon either side of the tip. Length, ♂ .8 in.; ♀ .86 in. Breadth across third abdominal segment, ♂ .38 in.; ♀ .36 in. Breadth across mesothorax, ♂ .34 in.; ♀ .30 in. Depth in middle of abdomen, ♂ .13 in.; ♀ .17 in.

♂ Draper's Valley, Virginia, (H. E. Scudder.) ♀ N. Y., (Uhler.) Penn., (H. Coll.) 3 specimens.

PYCNOSCELUS, Nov. gen. (πυκνὸς, σκέλος.)

A genus allied to the preceding, the males of which are wingless. I have no specimens of the female.

Head as in *Cryptocercus*, but proportionally larger; thoracic segments, and especially prothorax, very much arched, so as to form nearly a semicircle; prothorax only a little broader than long; the hind edge straight; the edge of the front and sides as well as the sides of the meso- and metathorax turned upwards very slightly, forming a delicate rim; the hind border of the meso- and metathorax is curved backwards a little at the outer angle; wings entirely wanting; the abdomen is very much expanded and flattened to a thin sheet posteriorly, so as to show no arching whatever; the segments narrow very rapidly towards the extremity, so that the hind curve is very broad and regular; the abdomen is half as broad again as long, and only equals in length the pro- and mesothorax together; hind edge of the seventh segment straight, but curved backwards somewhat at the outer angle; segments nearly uniform in length along the median line. The eyes are of moderate size, uniform with the surface of the head, nearer to one another than in *Cryptocercus*, pyriform, the broadest portions toward one another, the longitudinal diameter lying in the plane of the head; the antennæ are slender, not exceeding half the length of the body; the first joint not large, swollen at tip, and twice as long as the second, which with the succeeding is cylindrical; the third is as long as the first, and equals the succeeding four in size; all the joints are covered with very short hairs; the thoracic segments much hollowed below, giving space for the legs, which may be packed so closely as not to appear above the plane which unites the edges of the segments. Legs similar in every respect to those of *Cryptocercus*, except that the tibiæ are more flattened; and there is a distinct though small pad between the claws. The supraänal and the subgenital plates, which are exactly similar in character, are thrice as broad as long, regularly curved behind, with a slight fissuration in the middle, making them slightly bilobed; the cerci are very small but stout, pointed, flattened, with a medial ridge, nearly as broad at base as long, smooth, without trace of annulation; the styles are slender, cylindrical, bluntly pointed, of about the length of the cerci and inserted just within them.

* 1. P. OBSCURUS, nov. sp.

Of an uniform reddish brown above and below, shining upon the whole surface with the exception of the upper surface of the abdominal segments succeeding the third; head, thorax, and first three abdominal segments above with minute and distant punctures; upon the dulled surface these are exchanged for minute raised points, irregularly scattered over the general surface, becoming larger and bead-like upon the posterior borders of the segments, where they are equi-distant; these last are also found upon the posterior border of the third segment; a very faintly impressed median line upon the prothorax; the eyes are black; the antennæ dusky brown; the parts of the mouth yellowish brown; the legs the same, with a reddish tinge; the segments of the abdomen beneath have no punctures. Length, .46 in.; length of abdomen, .19 in. Breadth at hind border of prothorax, .19 in.; breadth at fourth abdominal segment, .29 in.; length of cerci, .015 in.

Greenfield, Mass., (Sanborn.) " In woods, under stones." 1 specimen.

PHASMIDA, Leach.

DIAPHEROMERA, GRAY.

*1. D. FEMORATA.

Spectrum femoratum, Say, App. Long's Second Expedition ; 297. (1824.)
"   "   Say, Am. Ent.; III. pl. 37. (1828.)
"   "   Say, Ent. of N. Am. (Ed. Le Conte.)   I. 82, 197. (1859.)
"   "   Harris, Cat. Ins. Mass.; 56. (1835.)
Diapheromera Sayi, Gray, Synopsis of Phasmidæ; 18. (1835.)
"   "   Serville, Hist. Nat. d. Orth.; 247. (1839.)
"   "   Charp., Orth. Descr. et Depict.; Pl. 6. (1841.)
"   "   Westw., Brit. Mus. Cat., Orth., Part I. Phasmidæ; 20. (1859.)
Bacteria (Bacunculus) Sayi, Burm., Handb. d. Ent.; II. 566. (1838.)
"   "   "   Burm., Zeitschr. f. Ent.; II. 39. (1840.)

Bacteria femorata, Haldeman in Icon. Encyc. (1857.)
Bacunculus femoratus, Uhler in Harris, Inj. Ins., 3d ed.;
146. (1861.)
Mass., (H. Coll., Sanborn, S. H. S.) N. H., (H. Coll.)
Illinois, (Uhler.) Red River Settlements, British America,
(S. H. S.) Nebraska, (Mus. Comp. Zoöl.)

GRYLLIDES, Latreille.

TRIDACTYLUS, OLIVIER.

The fact that this name is inapplicable in its signification to some of the species contained in it, is no valid reason for its disuse; and even should it be so considered, the name *Heteropus* proposed as early as 1805 by Palisot de Beauvais, must supersede that of *Xya*, so commonly in use, which was not proposed till four years later. If, however, the species having but two of the slender appendages at the termination of the posterior tibiæ should be found to differ generically from those having three, *Tridactylus* would have to be retained for the latter, with *Heteropus* as a synonym; and *Xya* should be applied to the former. That this may yet be found to be the truth, is indicated still further by the fact that the three-fingered species found in this country have a peculiar conformation of the anterior tibiæ, which, however, is a feature of the structure of the male alone, — a fact for which I am indebted to the scrutiny and kind communication of Mr. Uhler; this peculiarity is a lateral projection of an aduncate appendage inwards at the base, to the extremity of which, beside the hook, the tarsi are attached. As a figure will explain much better than any description I can give the form of these parts, I have drawn the anterior tibiæ and tarsi of *T. apicalis*, Say ♂. In the two species (*T. apicalis*, Say, and *T. terminalis*, Uhl.) which show this character, a further difference is to be seen between the males and females, in the more

Fig. 1. swollen prothorax of the former.

1. T. APICALIS, Say, Journ. Acad. Nat. Sc. Phil.;
IV. 310. (Fig. 1.) (1825.)
T. apicalis, Say, Ent. of N. Am. (ed. Le Conte);
II. 239. (1859.)
Xya apicalis, Burm., Handb. d. Ent.; II. 742. (1838.)

This is our largest species, the length of the body being fully one third of an inch in both sexes, and of a lighter color than the other species. I suspect that Say had specimens both of this and *T. terminalis*, Uhl., and confounded them together; his description applies best to this, while he endeavored to include them both when he said, "length, more than one fifth of an inch."

Alabama, Prof. Hentz, (H. Coll.) Kentucky, Mr. J. P. Wild, (Uhler.) 2 ♂, 1 ♀.

*2. T. TERMINALIS, Uhler Mss.

This species is darker than *T. apicalis*, the head and thorax being sometimes pitchy black, at others showing upon these parts reddish brown spots similarly disposed to those upon *T. apicalis*; — the two broad transverse fasciæ and the terminal spot upon the outside of posterior femora, which are only faintly indicated in *T. apicalis*, are here black and very distinct; the wings reach only the tip of the abdomen, while in the males of *T. apicalis* they extend considerably beyond it; it is a much smaller species than the preceding, and occupies a more northern area. Length from .25 to .30 inches.

Cambridge, Mass., May 20, (H. Coll.) Maryland, (Uhler.) So. Illinois, (Thomas, Uhler.) 1 ♂, 6 ♀.

3. T. MINUTUS, nov. sp.

This species resembles *T. terminalis* very much in its markings and coloration; but it is in general darker, and the markings are more distinct. The middle femora and tibiæ, and the posterior femora are very dark with narrow white bands, sometimes broken; the segments of the abdomen are bordered with white posteriorly; the wings in the only mature individual I have seen extend a little beyond the extremity of the abdomen. There are but two terminal appendages of the posterior tibiæ; the males want the aduncate lateral appendage of the anterior tibiæ. Length, .14 to .16 inches.

So. Illinois, (Thomas, Uhler.) 4 ♂.

GRYLLOTALPA, LATREILLE.

Curtilla, Oken, Lehrbuch d. Naturgesch.; III. I. 445 (1815.)

*1. G. BOREALIS, Burm., Handb. d. Ent.; II. 740. (1838.)

G. brevipennis, Serv., Hist. Nat. d. Orth.; 308. (1839.)

" " Harr., Report, 3d ed.; 149, fig. 68. (1862.)

G. Americana, Say Mss., and Harr., Cat. Ins. Mass.; 56. (1835.)

"Sides of pond, burrowing in moist earth, June, July, Sept." — H.

Mass., (H. Coll., Sanborn.) Island of Nantucket, (Mus. Comp. Zoöl.) Vermont, (Mus. Comp. Zoöl.)

*2. G. LONGIPENNIS, nov. sp.

Figured in Catesby, Nat. Hist. N. Car.; I. pl. 8.

This species does not differ in any respect from *G. borealis*, save in the greater size and comparatively greater breadth of the wing-covers, which cover rather more than half the abdomen, and in the much greater length of the wings, which extend considerably beyond the extremity of the abdomen; there is a very slight difference in some of the prominences of the anterior trochanters; in coloration and general proportions and absolute size it does not differ from the preceding; it has much the general appearance of *G. hexadactyla*, Perty, from which it differs in being somewhat stouter, in having the teeth of the anterior tarsi long and slender, and in that the ocelli are oval and not subrotund. Length, 1.33 inches.

Mass., (H. Coll.) Maryland, (Uhler.) 2 specimens.

GRYLLUS, LINNÆUS.

*1. G. LUCTUOSUS, Serv., Hist. Nat. d. Orthop.; 335. (1839.)

This species is readily distinguished from all other N. England species, by the great length of the wings, which, surpassing in length the wing-covers, hang over the extremity of the abdomen; contrary to the supposition of Serville, this is true of the male as well as of the female.

Length of ovipositor in seven individuals, .29–.39 in., average .35 in., length of hind femora, .20–.245 in., average, .225 in.

Mass., (Agassiz, Shurtleff, S. H. S.) Cape Cod, (Sanborn.) N. Hampshire, (Miss Edmands.)

*2. G. ABBREVIATUS, Serv., Hist. Nat. d. Orth.; 336. (1839.)
Acheta tripunctata, Harr., Cat. Ins. Mass.; 56. (1835.)
Acheta abbreviata, Harr., Report, 3d ed.; 152. fig. 69. (1862.)

This and *G. luctuosus*, Serv. are our largest species; it is remarkable for the great length of the ovipositor of the female; the wing-covers are generally dark, bordered with light brown, though sometimes they are quite pale throughout; they generally quite cover the abdomen. I have never seen this species north of Cape Cod, and am inclined to think it a more southern form, the insect fauna of Cape Cod being closely allied in character to that of Pennsylvania.

Length of ovipositor in seventeen individuals, .34–.42 in., average, .37 in.; length of hind femora, .21–.23 in., average .22 in.

Mass., (H. Coll., Agassiz.) Cape Cod, (Sanborn, S. H. S.) Maryland, (Uhler.) The specimen marked *A. abbreviata* ♂ in Harris's collection belongs to his *A. nigra*.

*3. G. ANGUSTUS, nov. sp.

This species is most closely allied to *G. abbreviatus*,

Serv., but differs from it very distinguishably in its much greater slenderness; in the relative proportions of the length of hind femora and ovipositor of the females, it agrees with *G. abbreviatus*, but the latter is a heavy, clumsy species; in *G. abbreviatus* the breadth of the body is contained in the length about three times; in *G. angustus* about four times; in coloration it does not differ from *G. abbreviatus*; it seems to be quite a rare species.

Length of ovipositor .30–.34 in., average, .32 in.; length of hind femora, .18–.20 in., average, .19 in.

Cambridge, Mass., and Cape Cod, (S. H. S.) 3 ♀.

*4. G. NEGLECTUS, nov. sp.

This is our most common species, and probably the one which Harris intended to describe in mentioning *G. nigra*, but the specimens in his cabinet marked " unique " show his *nigra* to be another species. The head, thorax, and body, as well as the hind femora, are pitchy black, the elytra of both ♂ and ♀ are dark, sometimes jet black, but frequently of quite a light ochraceous brown; indeed, the elytra of almost all our species vary to this extent in coloration; the elytra of the females generally cover about two thirds of the abdomen, although sometimes they entirely conceal it; those of the males extend to the extremity of the abdomen; the ovipositor in this species is proportionally shorter than in either of the preceding species, and is also a smaller species than any of the preceding.

Length averaging a little more than half an inch; length of ovipositor in nine individuals .23–.32 in., average .28 in.; length of hind femora, .16–.21 in., average, .20 in.

Mass., (Mus. Comp. Zoöl., Miss Edmands, Sanborn, S. H. S.) Cape Cod, (Sanborn, S. H. S.)

*5. G. NIGER.

Acheta nigra, Harr., Report, 3d ed.; 152. (1862.)

This species agrees in size with *G. neglectus*, but differs from it in the much shorter ovipositor, which is shorter in this species than in any other. A single female is the only remaining type of Harris's *nigra*. The specimen marked *A. abbreviata* ♂ in the Harris collection, is a ♂ of this species; it does not differ materially from *G. neglectus* in coloration, being generally not so dark, with more or less reddish hind femora, while it has the wing-covers somewhat longer than there; it seems to be our rarest species.

Length of ovipositor in two individuals, .23–.26 in., average, .245 in.; length of hind femora, .21–.23 in., average .22 in. 2 ♂, 2 ♀.

Mass., (H. Coll., S. H. S.)

Neither of these two latter species can be referred to *G. Pennsylvanicus*, Burm., which I have received from Maryland through Mr. Uhler. *G. Pennsylvanicus* agrees with *G. nigra* in the length of the ovipositor of the female, but it differs from it in the shortness and form of the wing-covers, as well as in their color and that of the legs, in which characters it agrees well with *G. neglectus*, with which it cannot be associated on account of the shortness of the ovipositor. I have another species, smaller than any here mentioned, which I took in Minnesota, which has a very long ovipositor, and also some quite peculiar forms from southern Illinois, received from Mr. Thomas, but I reserve their description till the reception of more specimens. The males of all these species are very difficult to distinguish; indeed, it cannot be done satisfactorily without a large number of specimens; — the study of this genus is certainly a very perplexing one.

NEMOBIUS, SERVILLE.

1. N. EXIGUUS.

Acheta exigua, Say, Journ. Ac. Nat. Sc. Phil.; IV. 309. (1825.)

" " Say, Ent. of N. America (Ed. Le Conte); I. 238. (1859.)

St. Louis, Missouri, (S. H. S.) Minnesota, (S. H. S.)
* 2. N. VITTATUS, Harr., Report, 3d ed.; 152.   (1862.)
Acheta servilis, Say Mss., and Harr., Cat. Ins. Mass.;
   56.   (1835.)
This species differs from the preceding in being of a darker color, and in having the hind femora somewhat ferruginous, besides that the wing-covers differ somewhat in their markings.

Mass., (H. Coll., Miss Edmands, Sanborn, S. H. S.) Maine, (Packard.) Conn., (S. H. S.)

* 3. N. FASCIATUS.
Gryllus fasciatus, De Geer, Mem.; III. 522, pl. 43,
   fig. 5.   (1773.)
Acheta hospes, Fabr., Syst. Ent.; 281.   (1775.)
   "   "   Fabr., Spec. Ins.; I. 355.   (1781.)
   "   "   Fabr., Mant. Ins.; I. 232.   (1787.)
   "   "   Fabr., Ent. Syst.; II. 32.   (1793.)
Gryllus hospes, Oliv., Enc. Meth.; IV. 636.   (1791.)

This species, except in the length of wings and wing-covers, is very similar to *N. exiguus;* it is somewhat larger, and has the terminal half of the last joint of maxillary palpi, black; the elytra in both sexes are light brown, with the veins darker, especially in the female, and extending to the extremity of the body; the wings in both sexes are as long again as the body, and the ovipositor of the female is so long as nearly to reach their tip. Length of ovipositor, .25 in. 4 specimens.

Mass., (Shurtleff, " flying against windows at night. Aug. 14.") Delphi, Indiana, (Mus. Comp. Zoöl.) Charleston, S. C., (Mus. Comp. Zoöl.)

I have but little doubt that this species is identical with De Geer's *G. fasciatus*, notwithstanding the somewhat greater size of his figure. But if I have been mistaken in this reference, I shall not have committed so grievous an error as Dr. Fitch has done (3d Report on Noxious Ins. of

N. Y. 132), in referring it to the genus *Œcanthus*, blaming Harris for a mistake never made by him, (see below, under *Œcanthus niveus*, p. 432,) who, he says, "was evidently unacquainted with the work of De Geer and the characters he assigns to these insects, or he would have been aware of his error, the marks by which this species is distinguished being so plain, and so explicitly stated by that author." How was it possible for so acute an observer as Dr. Fitch to overlook the close resemblance between De Geer's figure (save in the length of the wings), and our common *Nemobius vittatus*, Harr., and refer the species to *Œcanthus*, especially when *Œcanthi* of the same sex are figured upon the same plate! Moreover, is it not " explicitly stated by that author," in his description of *G. fasciatus*, that the posterior femora are stout and very wide, that the tibiæ of the same pair have long spines, and that the head and thorax are more hairy than ordinarily? Are these characters of *Œcanthus*, or of *Nemobius* and *Gryllus*? Burmeister has already intimated that the relations of this very species are with *Gryllus* rather than with *Œcanthus*. (Burm., Handb. II. 733.)

ŒCANTHUS, SERVILLE.

* 1. Œ. NIVEUS, Serv., Ann. Sc. Nat.; XXII. 135. (1831.)
Œcanthus niveus, Serv., Hist. Nat. d. Orth.; 361. (1839.)
"   "   Harr., Cat. Ins. Mass.; 56.   (1835.)
"   "   Harr., Report, 3d ed.; 154, figs. 71,
                   72.   (1862.)
"   "   Fitch, 3d Report, Noxious Ins. N. Y.;
                   131.   (1856.)
Gryllus niveus, De Geer, Mem.; III. 522, pl. 43, fig. 6.
                                           (1773.)
"   "   Oliv., Enc. Meth.; IV. 637.   (1791.)
Œcanthus cylindricus, Say Mss., and Harr., Cat. Ins.
                   Mass.; 56.   (1835.)

Œcanthus fasciatus, Fitch, 3d Report, Noxious Ins. N. Y.; 132 (omitting reference). (1856.)
Not only do individuals of this species differ from one another, as Fitch mentions in his distinctions between *Œ. niveus* and *Œ. fasciatus*, but even to a much greater extent, some males having three branches of the "fiddle-bow," and some even five, while they take their origin and termination at very different points in different individuals, and vary besides very much in coloration, many individuals being met with of quite a dark color, especially upon the abdomen and legs. These all belong, however, to one species, no differences being discoverable upon which true natural groups can be founded; nor are there any such concomitant characters among them as Fitch asserts; neither did Harris make any such blunder as to have misunderstood the sexes in this genus as is alleged by him. It would be strange indeed if an entomologist were not acquainted with the very apparent differences existing between them in the prolonged ovipositor of the female, and peculiar structure of the wing-covers in the male.

Mass., (H. Coll., Miss Edmands, Sanborn, Shurtleff, S. H. S.) Conn., (Norton, S. H. S.)

I have never met with *Gryllus bipunctatus*, De Geer, which Fitch refers to under the name of *Œ. punctulatus*. (3d Rep. 133).

LOCUSTARIÆ, Latreille.

In a paper on the genus *Raphidophora*, (Proc. Bost. Soc. Nat. Hist.; VIII. 6,) I enumerated the known species inhabiting the United States, amounting in all to three, and added to them descriptions of four others. Since then I have had an opportunity of examining many other species from various parts of the country, through the kindness of Mr. Uhler, and have made a study of others in the Museum of Comparative Zoölogy, so that the number of species is found to be very considerable, and to form quite a distinct feature of the North American Orthopteran fauna. I was able to show there that *R. xanthostoma*, Scudd. should not be united in the same genus with the others, and that *R. subterranea*, Scudd. had features in its structure which lacked conformity with those possessed by other members of the genus. A closer and more extended study has convinced me that there are here three well-marked genera, and that no one of them can properly be referred to *Rhaphidophora*. *Ceuthophilus* may be applied to the more abundant forms, living in concealment under stones, with which must be associated *R. stygia*, Scudd., found in the shallow Hickman's Cave; of this genus I am acquainted with no less than twelve species, besides one which I have not seen, *Phal. scabripes*, Hald., undoubtedly belonging here. Under the genus *Hadenœcus* we may place *R. subterranea*, Scudd., restricted to the deep caves of Kentucky, while *Tropidischia* is proposed for the genus under which *R. xanthostoma*, Scudd. of California should be placed.

CEUTHOPHILUS, Nov. gen. (κεῦθος, φίλος).

Head rather large, oval; antennæ long, slender, cylindrical; first joint as broad as long, larger and stouter than the rest, which are about equal in thickness, gradually tapering to the extremity; second quite short; third longest; the remainder unequal. Eyes subpyriform, subglobose, crowded against the first swollen joint of antennæ. Maxillary palpi long and slender; first two joints equal, smallest; third fully equal in length to first and second together; fourth three-fourths as long as third; fifth nearly as long as third and fourth together, somewhat curved, swollen towards extremity, split on the under side almost its entire length. Sides of the thoracic nota broad, mostly concealing the epimera; wings wanting; legs rather long; coxæ carinated externally, the third pair but slightly, the first pair having the carina elevated into a sharp, the second into a dull, point at the middle; first two pair of femora mostly wanting spines; hind femora thick and heavy, turned inward at the base, channelled beneath. Ovipositor generally rather long, nearly straight, but a little concave above, rounded off somewhat abruptly at the extremity to the sharp, upturned point.

This genus differs from *Rhapidophora* in the much shorter legs, in the comparative length of the joints of the maxillary palpi, in wanting the terminal spines of the first two pair of femora, and the unusual development of the terminal spines of posterior tibiæ, as also those upon the first tarsal joints, in the non-compressed joints of the tarsi, and the shortness of the cerci.

*1. C. MACULATUS.

Ephippigera maculata, Say Mss., and Harr. Cat. Ins. Mass.; 56. (1835.)
Raphidophora maculata, Harr., Report, 1st ed.; 126. (1841.)
Phalangopsis maculata, Harr., Report, 3d ed.; 155, fig. 73. (1862.)
Phalangopsis lapidicola, Burm., teste Erichson, Archiv. f. Nat.; 9.227, (see No. 3.) (1843.)
Raphidophora lapidicola, (pars) Scudd., Proc. Bost. Soc. Nat. Hist.; VIII. 7. (1861.)
(Not *Phalangopsis lapidicola*, Burm.)

This species has the posterior tibiæ of the male waved at the base in mature individuals,—which is true of this species only.

Mass., (H. Coll., Agassiz, Shurtleff, Sanborn, S. H. S.) Vermont, (Mus. Comp. Zoöl.) Norway, Maine, (Verrill.) Cape Elizabeth, Maine, (Morse.) Anticosti, Gulf of St. Lawrence, (Verrill.)

*2. C. BREVIPES, nov. sp.

A species very closely allied to the preceding, but of a

smaller size, and differing from it in its markings and proportions. It is of a pale, dull, brown color, very profusely spotted with dirty white spots, not so large or so frequently confluent as in *C. maculatus*, except near the extremity of the hind femora, where they nearly form an annulation. The mottling of the pronotum is somewhat different than in *C. maculatus;* the hind legs are proportionably shorter, as is also the ovipositor, the spines of whose inner valves are duller.

Length scarcely more than half an inch; average length of hind femora, .44 inch; average length of ovipositor, .25 inch; 2 ♂ 6 ♀.

Grand Manan Is., Maine, (Verrill.)

3. C. LAPIDICOLUS.

Phalangopsis lapidicola, Burm., Handb. d. Ent.; II. 723.
(1838.)
Raphidophora lapidicola, Burm., Germ. Zeitsch. f. Ent.;
II. 72.   (1840.)
"   "   (pars) Scudd., Proc. Bost. Soc.
Nat. Hist.; VIII. 7.   (1861.)

This species is very closely allied to *C. maculatus*, differing from it in style of mottling of the upper surface, and in that the males do not have the posterior tibiæ waved. Since there are two species, (this and the following,) which both correspond to the description of Burmeister's *lapidicola*, I have chosen to apply his name to that one, of which a specimen is to be found in the Cambridge Museum, labelled thus by Mr. Haldeman some years since. I had not seen it or any southern species previous to the publication of my paper on *Raphidophora*. 4 ♂ 3 ♀.

Maryland, (Uhler.) Pennsylvania, (Mus. Comp. Zoöl.) Georgia, (Mus. Comp. Zoöl.)

4. C. UHLERI, nov. sp.

This species also is closely allied to all the preceding,

but especially to *C. lapidicolus*, from which, however, it differs in its markings more than *C. lapidicolus* does from *C. maculatus*. The ground-color is reddish brown, and the spots which make up the mottling are distributed more regularly than in the preceding. It differs from it further in the greater length of the antennæ, and in the presence of spines upon the under side of the hind femora; these are spined both upon the inner and outer edge, those of the inner edge being minute, regularly arranged, and of equal size, while those of the outer edge to the number of 5–8 only, are much larger, longer, of unequal length, and irregularly arranged. As the only female I have seen wants the hind femora, I cannot tell whether the males and females differ in the character of these spines as is the case in some species. The hind legs of this species are proportionally longer than in any previously mentioned.

Average length of body .65 inches; average length hind femora .70 inches; length of antennæ about 1½ inches; length of ovipositor .35. 3 ♂ 1 ♀.

Maryland, (Uhler.)

5. C. SCABRIPES.

Phalangopsis scabripes, Hald., Proc. Ac. Nat. Sc. Phil.;
VI. 364. (1853.)
Raphidophora scabripes, Scudd., Proc. Bost. Soc. Nat.
Hist.; VIII. 7. (1861.)

This is the only described species of the genus which I have not seen. The darker portions of the hind femora of all the species have scabrous surfaces.

Alabama, (*teste* Haldeman.)

6. C. DIVERGENS, nov. sp.

A species recalling *C. lapidicola* by its coloration and markings, which in general appearance it much resembles, but from which, as from all others I know, it may be distinguished by the peculiar disposition of the spines upon the posterior tibiæ, which, in addition to the row of minute crowded spines directed downward which all have upon either edge of the under-side, have also five spines of a peculiar character placed in each of these rows; they are quite long, placed at regular distances from one another, from the tip of the tibiæ to near its base, those upon either row alternating with one another, and directed in almost exactly opposite directions; they do not point backwards at the same angle with the others, but are turned outwards nearly at right angles to the tibiæ; the tibiæ of the other legs also partake of this character to some extent; — in this species the spines of the posterior femora are altogether wanting in the female, while the male has spines similar to those of *C. Uhleri.* Antennæ quite long; hind femora in male stouter than in female.

Length of body, .45–.60 in.; hind femora, .25–.30 in.; antennæ, 1.5–2 inches. 1 ♂ 2 ♀.

Nebraska, (Mus. Comp. Zoöl.)

7. C. LATENS, nov. sp.

Pale yellowish brown, with darker streaks upon the hind femora and two broad bands of dark reddish brown along the whole dorsum, extending half-way down the sides, dotted irregularly with brownish yellow spots most profusely on the abdomen, and separated from one another by a narrow, irregular band of the same color; head above, reddish brown; below, yellowish brown; tips of the femora dark; no spines upon the under-side of the hind femora. The hind femora are thick and stout, and the whole hind leg shorter than in most of the species. The ovipositor is shorter than usual in this species, though not nearly so short as in *C. californianus.*

Length of body, .65 in.; of hind femora, .5 in.; of ovipositor, .27 in. 1 ♀ (antennæ broken).

Illinois, (Uhler.)

8. C. NIGER, nov. sp.

Most nearly allied to *C. latens* by the shortness of the

hind-legs and of the ovipositor. It is, however, wholly of a black color with a reddish tinge, especially about the head, under-surface of body, hind femora, and ovipositor; the spines of the legs are all reddish brown; there are also some traces of a reddish tinge upon the pronotum, which thus exhibit obsolete vestiges of the peculiar markings of the *Ceuthophili* hitherto mentioned; the hind femora are unusually slender though short; the claws of all the feet are twice as long as ordinarily, and the denticulations of the inner valves of the ovipositor are very slender and long; the antennæ are quite short; the eyes are more nearly circular and more globose than usual, and do not hug the base of the antennæ so closely.

Length of body, .6 in.; of hind femora, .4 in.; of ovipositor, .27; of antennæ, 1 in. 1 ♀.

Rock Island, Illinois, (Uhler.)

9. C. CALIFORNIANUS, nov. sp.

Fuscous, paler beneath and upon the front of the head; a narrow pale median line; eyes black; palpi pale; antennæ light brown. The hind femora are short and thick, and have no spines upon the under-surface; the claws of all the tarsi are quite long; the ovipositor is remarkably short, no longer than the cerci; the denticulations of the inner valves are rather prominent but dull.

Length of body, .53 in.; of hind femora, .22 in.; of ovipositor, .09 in. 1 ♀.

San Francisco, Cal. (Mus. Comp. Zoöl.)

I have a species from Texas, apparently more closely allied to this than to any other, though with an ovipositor of ordinary length, but in too mutilated a condition for description.

10. C. STYGIUS.

Raphidophora stygia, Scudd., Proc. Bost. Soc. Nat. Hist.;
VIII. 9. (1861.)

Hickman's Cave, Kentucky, (Hyatt.)

11. C. Agassizii.
Raphidophora Agassizii, Scudd., Proc. Bost. Soc. Nat.
Hist.; VIII. 11.   (1861).
Gulf of Georgia, Washington Territory, (A. Agassiz.)

12. C. gracilipes.
Phalangopsis gracilipes, Hald., Proc. Am. Ass. Adv.
Sc.; II. 346.   (1850.)
Raphidophora gracilipes, Scudd., Proc. Bost. Soc. Nat.
Hist.; VIII. 7.   (1861.)

This has longer legs than any other species, unless it be *C. stygia*, which it much resembles in style of marking and length of antennæ; this, together with the two preceding species, frequently have little suppressed spines upon the inner edge of the upper posterior half of the hind femora.

S. Illinois, (Uhler.)   N. York, (Uhler.)   Schooley's Mt., New Jersey, (Mus. Comp. Zoöl.) 3 ♂.

The first eight species of *Ceuthophilus* mentioned agree together remarkably in the distribution of the markings of the dorsum, as do also the three last among themselves, while *C. californianus* and the undescribed Texan species form a third distinct group.

HADENŒCUS, Nov. gen. (ᾅδης, ἔνοικος).

Body small and slender; head similar to *Ceuthophilus*; antennæ, very long and slender, exceeding the length of the body many times; basal joints much as in *Ceuthophilus*, except that the second is broader, and the fourth is more than half the length of the third; eyes as in *Ceuthophilus*; maxillary palpi very long and slender; first joint short; second fully twice as long as first; third quite long, fully equalling twice the length of second; fourth nearly as long as the third, slender at the base, thickened towards the tip; fifth, longer than third, of a similar form to the fourth, but more incrassated at the tip, compressed laterally, slightly curved, and split on the underside only at the tip. Tubercle of the vertex very small, pointed, bilobed. Epimera of the meso- and metathorax not covered by the sides of the meso- and metanotum; wings wanting; metasternum with a short sharp spine; legs remarkably long and slender; coxæ carinated externally, the first pair having the carina elevated in the middle to a point; femora without spines; hind femora turned inwards and a little swollen at the base, extending over only the basal half; under-surface delicately channelled; the two anterior tibiæ are slightly longer than their corresponding femora; tarsi much compressed laterally; anal cerci long and slender; ovipositor long and slightly ensiform, rounded off very gradually at the extremity to a delicate point.

This genus differs from *Rhaphidophora* in the proportional lengths of the joints of the maxillary palpi, in the want of spines on the first two pair of femora, and the peculiarity of character in those of the posterior tibiæ and basal tarsal joint, as well as the shape of the latter, in the convexity of the eyes, and in the non-development of spines on the coxæ of the mesothoracic legs; most probably *Raphid. palpata*, Charp., of Europe, belongs to this genus.

1. H. SUBTERRANEA.

Raphidophora subterranea. Scudd., Proc. Bost. Soc. Nat. Hist.; VIII. 8. (1861.)

Mammoth Cave, Kentucky, (Hyatt, D. C. Scudder.)

TROPIDISCHIA, Nov. gen. (τρόπις, ἰσχία).

Head similar to *Ceuthophilus*; antennæ long and slender, about three times the length of the body; first joint large and stout, considerably longer than broad; second much smaller but broader than the succeeding; third long and slender, narrowing anteriorly, the rest unequal; eyes subovate, very globose, slightly removed from the basal joint of antennæ; maxillary palpi, long and slender; first and second joints short, the second a little the longer; third more than twice the length of the second; fourth nearly as long as third; fifth nearly equal to third and fourth together, a very little curved, swollen at the tip, and split a little way down the under-side; tubercle of the vertex small, but sharply prominent, deeply bisected; sides of the thoracic nota shorter than in *Hadenœcus*, the meso- and metanotum not extending downwards so far as the pronotum; wings wanting; legs long and slender, especially the hindmost pair; the coxæ have the lower edge produced on the inner side to a small dull spine, and they are also carinated externally, the carinæ of the pro- and mesothorax being produced to a spine as in *Ceuthophilus*; the femora and tibiæ are four-sided, and have all the edges minutely and closely spined, except the posterior femora; these are swollen, though not heavily, at the basal portion, which is not turned inwards, as in the two preceding genera, and has the rectangular spinous character of the other femora upon the terminal half, and even affecting the swollen portion; the under-surface is deeply and narrowly channelled; the two anterior tibiæ are somewhat longer than their corresponding femora; there are no heavy spines upon any portion of the legs except upon the extremity of the hind tibiæ where there is a pair of moderately long ones; tarsi much compressed laterally; anal cerci blunt, channelled interiorly.

This genus differs from *Raphidophora* in the character of its maxillary palpi, the absence of any peculiar development of spines upon the legs, in the shape of the joints of the tarsi, and the globosity of the eyes.

1. T. XANTHOSTOMA.

Raphidophora xanthostoma, Scudd., Proc. Bost. Soc. Nat.
Hist.; VIII. 12. (1861.)

Crescent City, Cal. (A. Agassiz.)

UDEOPSYLLA, Nov. gen. (οὐδας, ψύλλα).

This genus is to be placed between *Ceuthophilus* and *Daihinia*. The body is heavier and stouter than in *Ceuthophilus*, with a larger head; the form of the body is that of *Ceuthophilus;* antennæ as in *Daihinia;* first joint larger and stouter than the rest, as broad as long, compressed anteriorly; third joint twice as long as second; remainder unequal; eyes small, subpyriform, docked on the antennal border, globose; maxillary palpi rather long; first and second joints equal and small; third, more than equal to the preceding together; fourth, little more than half as long as third; fifth, a little longer than third, somewhat curved, split along the whole under side; as in *Ceuthophilus* the pro- meso- and metanota nearly conceal the epimera of the thoracic segments; coxæ differing but slightly from *Ceuthophilus;* hind femora very heavy, thick, and especially broad, but not so much so as in *Daihinia*, where, as in this genus, the whole limb is swollen, and not the basal portion only, as in the preceding genera; in the males the hind femora are spined beneath; the fore and middle femora are shorter and heavier here and in *Daihinia* than in *Ceuthophilus;* tarsi, with the first and fourth joints equal and longest; second and third equal and small, the second overlapping the third above; the ovipositor is rather short, thick at base, slender at apical half, terminating much as in *Ceuthophilus*.

Fig. 2.

This genus differs from *Daihinia* in the longer, more slender, less robust, and less spinous legs, in the somewhat more slender body and smaller head, in the shorter maxillary palpi, and in the structure of the tarsal joints. See figs. 2, 3.

1. U. ROBUSTA.

Phalangopsis (*Daihinia*) *robustus*, Hald., Proc. Am. Ass. Adv. Sc.; II. 346.   (1850.)

Platte River above Fort Laramie, Nebraska, (Mus. Comp. Zoöl.) 2 ♂ 2 ♀.
I have examined Haldeman's types.

2. U. NIGRA, nov. sp. (Fig. 2.)

Shining black, with a faintly indicated, narrow, reddish dorsal line, a reddish tinge on the front of the face, the basal half of the inner surface of hind femora and the terminal half of the ovipositor, reddish. The hind femora of the male have, upon either edge of the under-surface, but especially on the inner, short but heavy spines, not crowded; the hind tibiæ are furnished on either edge of the upper surface with four or five opposite, long, and slender spines, between each two of which are placed three or four suppressed spines; there is a single row of short spines upon the under-surface, which become double towards the tip; the inner valves of the ovipositor have five teeth, growing longer and more curved towards the tip, where they are very long and slender.

Length of body, .8–.9 in.; of hind femora ♀ .56 in.; ♂ .68 in.; of ovipositor, .33 in.; of antennæ, about an inch. 1 ♂ 1 ♀.

Red River of the North, (Kennicott.) Minnesota, (S. H. S.)

DAIHINIA, HALDEMAN.

In this genus, while the tarsi of the mesothoracic legs are as they appear in allied genera, *the tarsal joints of the anterior and posterior pair are only three in number*, the first and last being of nearly equal length, with a *single* small joint between them, a very interesting exception to the almost universal rule among the *Locustariæ*. (See Fig. 3, *a. b.*)

Fig. 3.

1. D. BREVIPES, Hald., Proc. Am. Ass. Adv. Sc.; II. 346. (Fig. 3.) (1850.)

D. brevipes, Girard, Orthop. in Marcy's Expl. Red River of Louisiana; 246. Zoöl. Pl. XV. figs. 9–13. (1854.)

Platte River above Fort Laramie, Nebraska, (Mus. Comp. Zoöl.) 2 ♂, 1 ♀.
I have examined Haldeman's types.

CYRTOPHYLLUS, Burmeister.

* 1. C. CONCAVUS.
Pterophylla concava, Say Mss. and Harr., Enc. Am.;
      VIII. 42. (1831.)
" " Harr., Cat. Ins. Mass.; 56. (1835.)·
Platyphyllum concavum, Harr., Report, 3d ed.; 158, fig.
      74. (1862.)
Platyphyllum perspicillatum, Serv. *teste* Erichson, Archiv·
    f. Nat.; IX. 227. (see No. 2.) (1843.)
" " Uhl. in Harr., Report, 3d
    ed.; 158. (1862.)
(Not *Locusta perspicillata*, Fabr.)
Mass., (H. Coll., Agassiz.) Conn., (Norton.) N. Y., (Edwards, Akhurst.)

2. C. PERSPICILLATUS, Burm., Handb. d. Ent.; II. 697.
      (1838.)
Locusta perspicillata, Fabr., Spec. Ins.; I. 357. (1781.)
" " Fabr., Mant. Ins.; I. 233. (1787.)
" " Fabr., Ent. Syst.; II. 36. (1793.)
" " Stoll, Spectres, etc.; Pl. VII. a. fig.
    23. (1813.)
Platyphyllum perspicillatum, Serv., Hist. Nat. d. Orth.;
    445. (1839.)

This species differs from the northern one in its shorter, but equally broad wing covers, in the slightly broader sonorous apparatus of the male, and in the more robust legs. 1 ♂.
Texas, (Mus. Comp. Zoöl.)

PHYLLOPTERA, Serville.

* 1. P. OBLONGIFOLIA, Burm., Handb. d. Ent.; II. 693.
      (1838.)

Locusta oblongifolia, De Geer, Mem.; III. 445. Pl. 38, fig. 2. (1773.)
Gryllus oblongifolius, Harr., Cat. Ins. Mass.; 56. (1835.)
Phylloptera oblongifolia, Harr., Report, 3d ed.; 159. (omitting figure) (1862.)
Mass., (H. Coll., Agassiz, S. H. S.) 3 ♂, 2 ♀.

* 2. P. ROTUNDIFOLIA, nov. sp. (Fig. 4), figured as *P. oblongifolia*, Harris' Report, 3d ed., fig. 75.

This species agrees with the preceding in coloration in every respect, unless the color of the ovipositor of the female be different in fresh specimens. The wings and wing-covers are much shorter than in *P. oblongifolia*, the wing-covers, in consequence, being more ovoid. It differs from that species, also, in the shape of the prothorax, which, in *P. oblongifolia*, is much narrower at the anterior than at the posterior border, and has the angle formed by the deflexion of the sides quite sharp, while in *P. rotundifolia* the posterior border is scarcely wider than the anterior, and the angle of the sides is rounded. It is a smaller species than *P. oblongifolia*.

Length of body, .8 in.; of wing covers, 1 in.; of wings, (when closed,) 1.17 in.; of hind femora, .87 in.; of ovipositor, .37 in.

Mass., (Sanborn, Miss Edmands.) Vermont, (Mus. Comp. Zoöl.) Conn., (Norton.) Rhode Island, (H. Coll.) Illinois, (Mus. Comp. Zoöl.)

3. P. CAUDATA, nov. sp.

Similar in general appearance to *P. oblongifolia*, but having a larger body, with slightly longer wings, much longer legs, and a very long ovipositor. The prothorax is narrowed anteriorly, as in *P. oblongifolia*, while the lateral angles are rounded as in *P. rotundifolia*. The specimen I have examined is old and discolored, but faint tinges of green are left upon some parts, indicating that the general color was as in the preceding species.

Length of body, 1 in.; of wing covers, 1.5 in.; of wings, (when closed,) 1.8 in.; of hind femora, 1.4 in.; of ovipositor, .8 in. 1 ♀.
Texas, (Mus. Comp. Zoöl.)

MICROCENTRUM, Nov. gen. (μικρὸς, κέντρος).

Head oval, broader and stouter than in *Phylloptera*; tubercle of the vertex somewhat prominent, scarcely broader than first joint of antennæ, slightly furrowed; eyes broadly oval, very prominent; first joint of antennæ as broad as long, second one-third as large but also stout, remainder long and slender, cylindrical. Prothorax flat or very slightly concave above, anterior border very slightly concave, posterior quite convex; the sides nearly parallel, the length but little surpassing the breadth; lateral carinæ quite sharp; lobes of the side straight in front, well rounded and curving forwards behind, rounded beneath, deeper than broad; wing-covers with the triangular superior surface extending backwards farther than in *Phylloptera*, and the wing-covers themselves not regularly rounded as there, but with the inner border straighter till near the tip, the outer border sloped off toward the tip, and the tip itself more pointed (see figs. 4 and 5); legs slender, much shorter than in *Phylloptera*, especially the metathoracic; ovipositor very short, strongly curved, and bluntly pointed.

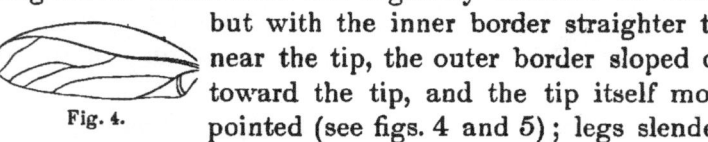
Fig. 4.

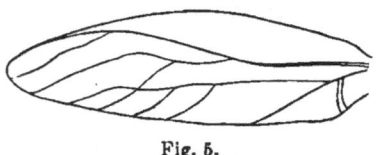
Fig. 5.

This genus differs from *Phylloptera*, to which it is most nearly allied, especially by the cut of the wing-covers and the shortness of the hind legs and ovipositor.

1. M. RETINERVIS.

Phylloptera retinervis, Burm., Handb. d. Ent.; II. 692.
(1838.)

of the North American Orthoptera. 447

Phylloptera curvicauda, Harr., Report, 3d ed.; 161, note.
(1862.)
(Not *Locusta curvicauda,* De Geer.)
North Carolina, (H. Coll.) Georgetown, D. C., (Mus. Comp. Zoöl.) 2 ♂.

*2. M. AFFILIATUM, nov. sp. (Fig. 5.)

This species is very closely allied to the preceding, but is a larger insect; the hind femora are proportionally shorter when compared with the wing-covers; the eyes are less prominent than there, and the tubercle of the vertex is slightly broader, with a broader and shorter medial furrow, forming rather a sort of shallow pit. A further distinction is seen in that the slightly hollowed front of the prothorax has a very small central tooth, which does not exist in *M. retinervis.*

Length of wing-covers, 1.75 in.; breadth, .56 in.; of hind femora, .9 in.; of ovipositor, .24 in. 4 ♂, 2 ♀.

Mass., (H. Coll., Miss Edmands.) Maryland, (Uhler.) Key West, (Mus. Comp. Zoöl.) Texas, (Mus. Comp. Zoöl.) Nebraska, (Mus. Comp. Zoöl.)

3. M. THORACICUM, nov. sp.

Locusta maxima viridis alis latissimis, Sloane., Nat. Hist. Jamaica; II. 201. Pl. 236, figs. 1, 2. (1725.)

Tubercle of the vertex rather prominent, narrow, faintly bilobed and divergent at the apex, with a narrow median groove; eyes as in *M. affiliatum,* but large; prothorax constricted anteriorly, the front border straight; side lobes broader and more amply rounded than in other species; lateral carinæ quite prominent, transversely ridged, raised at the posterior half quite considerably, and pinched where it is highest; hind border not so convex as in the preceding species, and slightly angulated; the top of the prothorax is hollowed, and has a faint medial and transverse furrow; wing-covers very closely and minutely punctured throughout; posterior tibiæ triquetral, expanded upon the upper surface, with a row of fine spines upon either side, a single row beneath; upper surface flat from tip to quite near the base, where it is suddenly rounded; ovipositor very small, slender, sharply upturned. The only specimen I have seen was preserved in alcohol, but still exhibits a short, median, longitudinal, blood-red streak on the face, and has the posterior third of the lateral carinæ of prothorax, with the whole hind border, edged with black.

Length of wing-cover, 2.2 in.; breadth, .8 in.; length of wings beyond wing-cover, .2 in.; of hind femora, 1.2 in.; of ovipositor, .25 in. 1 ♀.

Tortugas, Florida, (Mus. Comp. Zoöl.)

This species seems closely allied to *Phylloptera azteca*, Sauss.

PHANEROPTERA, Serville.

* 1. P. CURVICAUDA, Serv., Ann. Sc. Nat.; XXII. 159. (1831.)
Phaneroptera curvicauda, Burm., Handb. d. Ent.; II. 691. (1838.)
Locusta curvicauda, De Geer, Mem.; III. 446. Pl. 38, fig. 3. (1773.)
Gryllus locusta myrtifolius, Drury, Ill. Ex. Ent.; II. 78. Pl. 41, fig. 2. (1773.)
Gryllus (Phyllopterus) myrtifolius, Drury, Ill. Ex. Ent. (ed. Westw.); II. 88. Pl. 41, fig. 2, (omitting synonymy.) (1837.)
Pterophylla curvicauda, Harr., Cat. Ins. Mass.; 56. (1835.)
Phaneroptera augustifolia, Harr., Report, 3d ed.; 160. fig. 76. (1862.)

This species varies very much in size, and in the proportions of the length of the wing-cover to its breadth. I have seen specimens from Texas which apparently belong to this species.

Mass., (H. Coll., Agassiz, Shurtleff, Sanborn, S. H. S.)

Conn. (Norton.) Maine, (Packard.) Red River Settlements, British Am. (S. H. S.)

CONOCEPHALUS, THUNBERG.

*1. C. ENSIGER, Harr., Report, 3d ed.; 163. fig. 79.
(1862.)
C. ensiger, Harr., Cat. Ins. Mass.; 56.   (1835.)
Locusta acuminata, Stoll, Spectres; Pl. VIII. a. fig. 27.
(1813.)
(Not *Locusta acuminata*, Linn. and De Geer, nor *Locusta acuminata*, Fabr.)

Mass., (H. Coll., Sanborn, Shurtleff.) Cape Cod, (S. H. S.) Vermont, (H. Coll.) Conn., (Norton, S. H. S.) Illinois, (Mus. Comp. Zoöl.) Minnesota, (S. H. S.) Nebraska, (Mus. Comp. Zoöl.)

*2. C. ROBUSTUS, nov. sp.

Either pea-green or dirty brown; tubercle of the vertex tipped with black, not extending, or but very faintly and narrowly, down the sides; lateral carinæ of prothorax pale yellowish; wing-covers dotted with irregularly distributed black dots, most conspicuous in the brownish individuals. In form as in coloration, this species is much like *C. ensiger*. The shape of the conical projection of the vertex is the same, or a little stouter; it is a larger species, much broader and stouter than it, the wings broader, and when compared with the hind femora, a little longer than they are in *C. ensiger*; the spines upon the under side of the hind femora are larger than there, being noticed easily with the unassisted eye; the ovipositor of the female is much shorter than in *C. ensiger*, and finally the insect is much broader across the mesothorax, with a heavier sonorous apparatus in the male; wing-covers fully as long as the wings in the male; slightly longer than the wings in the female. The only difference between this species and *C. ensiger* in coloration is the usual lacking of the spots on the wing-cover in the latter, and in the same the presence of a broad black band on either side of the tubercle of the vertex, which exists in the former but seldom, and then it is very narrow.

♂ Length of wings, 1.7 in.; breadth in middle, .32 in.; of hind femora, .9 in. ♀ Length of wing-covers, 1.9 in.; extent of wing-covers beyond wings, .1 in.; breadth of wing-covers in middle, .22 in.; length of hind femora, 1 in.; of ovipositor, 1 in. 17 ♂ green, 19 ♂ brown, 1 ♀ green.

Cape Cod, by the sea-beach, Sept. (S. H. S.)

This peculiarity of having its individuals either wholly green or wholly brown, extends to *C. ensiger* also, and is found in both while the animals are alive; I have never seen a brown ♀ alive.

3. C. OBTUSUS, Burm., Handb. d. Ent.; II. 705. (1838.)
C. dissimilis, Serv., Hist. Nat. d. Orth.; 518. (1839.)
"   "   Harr., Report, 3d ed.; 164. (1862.)
Georgia, (Mus. Comp. Zoöl.) 7 ♀.

4. C. UNCINATUS, Harr., Report, 3d ed.; 164. (1862.)

The legs in this species are much shorter and more robust than in any other American species I have seen. Length of hind femora .65 in. I have seen only Harris's original specimen. 1 ♀.

Alabama, (H. Coll.)

5. C. CREPITANS, nov. sp.

The specimens I have examined were dried after an immersion in alcohol, and are uniform in their coloration throughout, with indications of light yellow lateral streaks on the prothorax, as in *C. robustus;* the tubercle of the vertex is devoid of markings; the shape of the tubercle is very similar to *C. robustus*, but it is stouter than there; head and prothorax punctured throughout, the latter with a faintly impressed median line. This is a large species, broad across the mesothorax as in *C. robustus*, having very long and narrow wings, long and slender legs, and a rather long ovipositor.

Length of body 1.5 in.; of wing-covers ♂ 2 in., ♀ 2.4 in.; breadth in middle ♂ .33 in., ♀ .28 in.; length of hind femora ♂ 1.1 in., ♀ 1.3 in.; of ovipositor 1.43 in.; of tubercle of vertex beyond the eye .12 in., 1 ♂, 2 ♀.

Texas, (Mus. Comp. Zoöl.) Nebraska, (Mus. Comp. Zoöl.)

XIPHIDIUM, SERVILLE.

* 1. X. FASCIATUM, Serv., Ann. Sc. Nat.; XXII. 159. (1831.)

Locusta fasciata, De Geer, Mem.; III. 458. Pl. 40, fig. 4. (1773.)

Pterophylla fasciata, Harr., Cat. Ins. Mass.; 56. (1835.)

Orchelimum gracile, Harr., Report, 3d ed.; 163. Fig. 78. (1862.)

Xiphidium fasciatum, Burm., Handb. d. Ent.; II. 708. (1838.)

Mass., (H. Coll., Shurtleff, Sanborn, Miss Edmands, Agassiz, S. H. S.) Maine, (Packard.) Vermont, (H. Coll.) Rhode Island, (H. Coll.) Conn., (Norton.) Cape Cod, (S. H. S.)

* 2. X. BREVIPENNIS, nov. sp.

Size of *X. fasciatum*, with which it agrees in coloration throughout, except that the wings are a little darker, the dorsal band is a little broader, and the ovipositor is of a reddish brown throughout, while in *X. fasciatum* it is green at the base; wings .08 in. shorter than the wing-covers; both shorter than the body; ovipositor nearly equalling the hind femora in length. In these respects it differs very much from *X. fasciatum*.

Length of body, .5 inch; of wing-covers, .33 in.; of hind femora, .43 in.; of ovipositor, .4 in.

Mass., (H. Coll., Agassiz, Miss Edmands, Shurtleff, Sanborn, S. H. S.) Cape Cod, (S. H. S.) Maine, (Packard.)

3. X. ENSIFER, nov. sp.

Green, with a reddish brown broad central stripe on top

of head and prothorax, narrowed in front, extending to tip of tubercle of vertex; wings very nearly of the same length as the wing-covers, both shorter than body, as in *X. brevipennis*; ovipositor very long, exceeding the length of the hindmost femora, of a brown color, deepest toward apex.

Length of body, .55 in.; of wing-covers, .30 in.; of hind femora, .48 in.; of ovipositor, .6 in. 2 ♀.

Lawn Ridge, Illinois, (Mus. Comp. Zoöl.)

ORCHELIMUM, SERVILLE.

* 1. O. VULGARE, Harr., Report, 3d ed.; 162. Fig. 77. (1862.)
Pterophylla agilis, Harr., Cat. Ins. Mass.; 56. (1835.)
Mass., (H. Coll., Sanborn, S. H. S.) Cape Cod, (S. H. S.) Conn., (Norton.)

* 2. O. CONCINNUM, nov. sp.

♂ Brownish green; a dark reddish brown dorsal streak upon the head and prothorax, becoming faint towards the hind border of prothorax, and narrowing anteriorly to the width of the tubercle of the vertex, passing over this down the front to the labrum, expanding broadly in the middle of the face; legs brownish green, tarsi dark brown, spines of tibiæ tipped with black; abdominal appendages reddish brown; wing-covers pellucid, veins grass-green, except the heavy transverse vein of the sonorous apparatus, which is brown; wings pale brownish green, extending a little beyond wing-covers; ♀ having the same markings as the ♂ except that all the nervures of the wing-covers are brown, and the wings are more dusky, and are shorter than the wing-covers; ovipositor reddish brown, a little curved and very pointed; a much slenderer and more graceful form than *O. vulgare*.

Length of body, .7 in.; of wing-covers, .84 in.; of wings beyond wing-covers, .08 in.; of hind femora, .6 in.; of ovipositor, .32 in. 1 ♂, 2 ♀.

Cape Cod, (S. H. S.)

*3. O. GLABERRIMUM.

Xiphidium glaberrimum, Burm., Handb. d. Ent.; II. 707.
(1838.)

The dorsal band here is bordered with black, as is also the outer edge of the sonorous apparatus of the male; antennæ very long; ovipositor slightly expanded in the middle.

Conn., (Norton.)   Georgia, (Gerhard.)   1 ♂, 1 ♀.

4. O. AGILE.

Locusta agilis, De Geer, Mem.; III. 457. Pl. 40, fig. 3.
(1773.)

Xiphidium agile, Burm., Handb. d. Ent.; II. 707. (1838.)

This species has a short ovipositor, shaped much as in *O. vulgare*, and a very narrow dark median streak down the face; it has a shorter pronotum than *O. vulgare*.

Maryland, (Uhler.)   Illinois, (Mus. Comp. Zoöl.) 2 ♀.

5. O. LONGIPENNIS, nov. sp.

Dorsal band on head dark reddish brown, broad, narrowing to the width of tubercle of vertex, not extending over the face, divided on the pronotum, forming two narrow bands, scarcely reaching either front or hind border; wing-covers very long and slender, a little shorter than the wings; tarsi with first two joints brown, the other two dark green; ovipositor not long, pointed, reddish brown; antennæ extending back beyond tip of wings.

Length of wing-covers, 1.16 in.; of wings beyond wing-covers, .1 in.; of hind femora, .72 in.; of ovipositor, .32 in. 1 ♀.

Texas, (Mus. Comp. Zoöl.).

THYREONOTUS, SERVILLE.

*1. T. PACHYMERUS.

Decticus pachymerus, Burm., Handb. d. Ent.; II. 712.
(1838.)

Conn., (Norton.) Mammoth Cave, Kentucky, (Hyatt.) 1 ♂, 2 ♀.

*2. T. DORSALIS.

Decticus dorsalis, Burm., Handb. d. Ent.; II. 713. (1838.) Mass., (Sanborn.) Rhode Island, (H. Coll.) Maryland, (Uhler.) 6 ♀.

Among other distinctions between these two species, it may be seen that *T. pachymerus* has the pronotum well rounded behind, while the hind margin of the other is nearly square; and the ovipositor is longer in *T. dorsale* than in *T. pachymerus*, as are also the hind legs.

ACRYDII, Latreille.

OPOMALA, SERVILLE (*emend.*)

*1. O. BRACHYPTERA, nov. sp.

Above reddish brown, dotted faintly with black, extending a little over the sides; sides dirty yellowish brown, with a faint dark streak extending from lower border of eye backwards over the lower border of pronotum; face dirty yellowish brown, dotted faintly with brownish spots; antennæ brown, darkest toward tip; legs reddish brown; tarsi darker, tibiæ with black tipped spines; hind femora with a row of black dots on upper edge, terminal lobe black; hind tibiæ at base and on under side toward the tip, black; wing covers yellowish brown; vertex of the head rather prominent, suddenly swollen in advance of the eyes, from thence sloping to a blunt rounded point, the edge upturned, and the median ridge prominent and sharp, becoming rounded on the head; wing-covers but little more than half the length of body; wings very short, nearly abortive.

Length of body, 1.05 in.; of antennæ, .46 in.; of vertex, .053 in.; of hind femora, .52 in.; of wing-covers, .42 in of wings, .1 in. 1 ♂.

Princeton, Mass. (S. H. S.).

CHLOËALTIS, HARRIS.

Chrysochraon, Fisch. Fr.

*1. C. CONSPERSA, Harr., Report, 3d ed.; 184. (1862.)
   C. abortiva,   "   "   "   "   (1862.)

Mass., (Sanborn.) New Hampshire, (H. Coll.) Eastern shore of Lake Winnipeg, British America, (S. H. S.) 5 specimens.

*2. C. VIRIDIS, nov. sp.

Vertex broad, expanding a little in advance of the eyes, beyond which the sides slope so as to form a right angle with each other, rounded at the apex; the edge upturned more or less; pronotum with the median and lateral carinæ parallel, distinct, sharp; wing-covers shorter than the body, a little longer than the wings.

♂ Top of head and prothorax green; sides of head and prothorax dirty brown, with an horizontal black band behind the eye, extending over prothorax; front of head yellowish brown; fore and hind legs reddish brown, mesothoracic legs green; spines of tibiæ tipped with black; wing-covers above green, upon the sides brown; body beneath yellowish. ♀ varying from olivaceous green to dark brown, with a dark band behind the eye as in the ♂; upon the top of the head a dark band extends from either side of the vertex, curving inwards and then outwards to midway between the median and lateral carinæ; hind tibiæ reddish brown.

Length of body, ♂ .6 in., ♀ 1 in.; of pronotum, ♂ .14 in.; ♀ .21 in.; breadth of pronotum, ♂ .07 in., ♀ .13 in.; length of hind femora, ♂ .4 in.; ♀ .6 in.; of wing-covers, ♂ .3 in., ♀ .42 in. 3 specimens.

Conn., (Norton.)

*3. C. PUNCTULATA, nov. sp.

Vertex broad, much as in *C. viridis*, but slightly more prominent; sides of the pronotum very nearly parallel, slightly divergent posteriorly; lateral and median carinæ distinct, sharp; wing-covers extending to tip of abdomen equally with the wings.

Vertex edged with reddish brown; a narrow reddish brown band extends along the lateral carinæ of pronotum to the eye, edged below with black; it extends also slightly upon the base of the wing-covers; abdomen, sternum, forelegs and mouth-parts, (except the black mandibles,) reddish brown; hind tibiæ yellowish brown, its spines tipped with black; all the tarsi darker; wing-covers green, with scattered small brownish spots.

Length of body, .95 in.; of pronotum, .19 in.; width of pronotum in middle, .09 in.; length of hind femora, .54 in.; of wing-covers, .7 in. 1 ♀.

Conn., (Norton.)

STENOBOTHRUS, FISCHER FR.

* 1. S. CURTIPENNIS.

Locusta curtipennis, Harr., Cat. Ins. Mass.; 56. (1835.)
Chloëaltis curtipennis, Harr., Report, 3d ed.; 184. Pl. 3, fig. 1. (1862.)

This is our most common species, and is very abundant. The figure in Harris's Report is a very poor one; the antennæ are more than one half too short, and the pronotum is inaccurately rendered.

Mass., (H. Coll., Agassiz, Miss Edmands, Shurtleff, Sanborn, S. H. S.) Maine, (Packard.) Conn., (Norton, S. H. S.) Red River Settlements, British America, (S. H. S.)

* 2. S. MELANOPLEURUS, nov. sp.

Vertex of the head broad, expanded to a blunt point on either side in front of the eyes, triangular, very blunt at the apex; edge upturned with a very slight median ridge, scarcely crossing the vertex; no foveolæ; pronotum with lateral carinæ nearly parallel, slightly approaching one another in the middle; median carina sharp, rather more distinct than the lateral; posterior border of pronotum straight; wing-covers slightly shorter than the body; wings nearly abortive.

Brown; sides of the pronotum, and of the first two or three abdominal segments, shining black; face and mouth-parts paler; a reddish brown, curved streak on the top of the head from inner edge of eye to lateral carinæ of pronotum; legs yellowish brown; posterior femora with one or two dark streaks on the sides; posterior tibiæ black at tip and base.

Length of body, .67 in.; of antennæ, .39 in.; of wing-covers, .38 in.; of hind femora, .42 in. 2 ♂.

Mass., (S. H. S.)   Maine, (Packard.)

Easily distinguished from any other species by the black sides of the pronotum.

* 3. S. LONGIPENNIS, nov. sp.

Vertex of the head as in *S. melanopleurus*, but with no median ridge, and having very distinct foveolæ, long, narrow, deep; lateral carinæ of pronotum equally prominent with the median, approximate, convergent anteriorly, divergent at posterior border; coarse, shallow punctures on the posterior half of pronotum; posterior border arcuated; wing-covers longer than body; wings scarcely shorter than wing-covers.

Head and thorax brown; a broad, black band on the sides, behind the eye, extending to hind edge of pronotum, limited above by the lateral carinæ, below merging into the brown; a narrow, straight, longitudinal streak on top of head, starting from inner border of the eye; parts of the mouth yellowish; antennæ yellowish brown at base, the rest brown or black; legs yellowish brown; hind tibiæ, except the black base, and slender portion of hind femora reddish brown, extremity black; abdomen yellow beneath, above brown; wing-covers uniform brown.

Length of body, ♂ .55 in., ♀ .7 in.; of antennæ, ♂ .36

in., ♀ .28 in.; of hind femora, .44 in.; of wing-covers, .65 in. 7 specimens.

Mass., (H. Coll., Miss Edmands, Sanborn, S. H. S.)

4. S. SPECIOSUS, nov. sp.

Vertex of the head quite broad, not -expanding at the sides, apex not rounded, the sides of the angle straight; edges upturned considerably; a slight median groove; no foveolæ; sides of the pronotum approximate, constricted in the middle; lateral foveolæ not so prominent and sharp as the median; wings as long as the wing-covers, extending beyond the tip of abdomen.

Above brown; below pale yellow; face yellowish brown; mouth-parts pale yellow; antennæ reddish brown; a narrow, curved streak on top of the head from inner edge of eye to lateral carinæ; a narrow, straight, white streak from eye to lateral carinæ; upper half of sides of pronotum brownish, darkest above; legs yellowish brown; spines of hind tibiæ tipped with black; wing-covers brownish at base, apical half pellucid, with rosaceous nervures; wings pellucid with rosaceous nervures; costa with a dark streak beyond the middle.

Length of body, .55 in.; of antennæ, .23 in.; of hind femora, .34 in.; expanse of wing-covers, 1 in.; of wings, .44 in. 1 ♂.

St. Paul, Minnesota, (S. H. S.)

The above-mentioned species have all immaculate wing-covers; those to be mentioned have all spots or streaks on them.

* 5. S. MACULIPENNIS, nov. sp.

Vertex of the head rather broad, with elevated edges, with no expansion of the sides in advance of the eyes; the apex blunt, with long, shallow foveolæ, broader toward the eye than at the apex; lateral carinæ of the pronotum convergent at the anterior half, very divergent at the posterior border, somewhat elevated but not rounded; median carina sharp, high; hind border of pronotum angulated, angle rounded; wings equally with wing-covers, extending back beyond the abdomen.

Head and top of pronotum green (some individuals brown); a reddish-brown, broad band behind the eye reaches the hind edge of sides of pronotum, limited above by the lateral carinæ, which are white, but crossing this posteriorly and filling the triangular space on top of pronotum, made by the divergence of the carinæ at the posterior border; sides of pronotum below the band brownish; wing-covers green, with a medial band of equidistant, square, black spots along its whole extent, besides a few irregularly scattered smaller black spots; sometimes the inner halves of the wing-covers are entirely of a rust-red color; legs yellowish brown; the hind femora sometimes streaked; the hind tibiæ plumbeous, with a broad, pale, indistinct annulation near the base; antennæ with the basal half reddish, the apical brownish.

Length of body, ♂ .55 in., ♀ .75 in.; of antennæ, ♀ .26 in.; of hind femora, .45 in.; of wing-covers, ♂ .4 in., ♀ .7 in.

Mass., (H. Coll., Shurtleff, Sanborn, S. H. S.)

*6. S. ÆQUALIS, nov. sp.

Very similar in general appearance to *S. maculipennis*, but a smaller species, with wing-covers extending only to the extremity of abdomen.

Vertex of head broad and blunt, the sides slightly swollen at the anterior border of the eye, the apex blunt, and all the angles rounded; the edge slightly and not sharply upturned; foveolæ shallow, short, nearly equiangular; lateral carinæ of the pronotum curved inward a little in the middle, not so prominent as the sharp medial carina; hind border slightly angular, but nearly straight; wings and wing-covers just reaching the extremity of the abdomen.

Colored much as in *S. maculipennis;* the lateral carinæ are yellow, and the band extends forwards to the eye; the band behind the eye is quite narrow, and the sides below it green, like the parts above, and the triangular dash of black upon the top of the pronotum at the hinder angles is much narrower than there, on account of the lesser divergence of the lateral carinæ; the medial band of the wing-covers containing the square black spots is yellowish brown as in *S. maculipennis*, and the extremity is pellucid.

Length of body, ♂ .45 in., ♀ .66 in.; of antennæ, ♂ .24 in., ♀ .2 in.; of hind femora, ♂ .3 in., ♀ .38 in.; of wing-cover, ♂ .34 in., ♀ .46 in.

Mass., (Shurtleff, Sanborn, S. H. S.) Maine, (Packard.) N. Y., (H. Coll.) Minnesota, (S. H. S.)

*7. S. BILINEATUS, nov. sp.

Vertex of the head slightly swollen at anterior border of the eye, but rounded as is also the apex; edges not upturned; foveolæ only indicated by a very faint, scarcely perceptible depression; sides of the pronotum rather suddenly constricted in the middle, the lateral carinæ broader than the sharp medial one; hind border of pronotum slightly rounded; wing-covers of the length of the abdomen.

Brown, generally dark; face yellowish brown, blotched with black; a narrow black band extends from the tip of the vertex along each side, extending back on to the pronotum by the side of the lateral carinæ to the posterior border, widening upon the pronotum; the usual black band behind the eye is rather broad, and there is another similar one on the pronotum farther down the side, with a yellowish streak between them; the wing-covers are brown with a medial row of darker spots; the legs are brownish, with the extremity of the hind femora and the base of the hind tibiæ darker.

Length of body, ♂ .5 in., ♀ .65 in.; of antennæ, ♂

.22 in., ♀ .19 in.; of hind femora, ♂ .32 in., ♀ .36 in.; of wing-covers, ♂ .36 in., ♀ .41 in.

Mass., (H. Coll., Sanborn, Shurtleff, S. H. S.)

*8. S. PROPINQUANS, nov. sp.

Very similar to the preceding, but of a larger size, and has proportionally longer wings. It is also closely allied to *S. maculipennis*.

Vertex broad, expanding but slightly at anterior border of the eye, the angle rounded, the apex blunt, edges scarcely if at all raised, foveolæ as in *S. maculipennis*, but more shallow; lateral carinæ of pronotum somewhat convergent in the middle, of equal prominence and sharpness with the medial; hind border of pronotum somewhat rounded; wings a little longer than abdomen.

Brown; band behind eye quite broad; lateral carinæ yellowish; a faint curved dark band from inner border of eye to lateral carinæ; antennæ yellowish brown; wing-covers brownish at base, transparent at tip, with a medial band of brown spots extending two thirds of the distance to the tip; legs yellowish brown; hind tibiæ plumbeous, with a broad pale annulation at base.

Length of body, ♂ .6 in., ♀ .75 in.; of antennæ, ♂ .26 in., ♀ .23 in.; of hind femora, ♂ .38 in., ♀ .48 in.; of wing-covers, ♂ .55 in., ♀ .68 in.

Conn., (Norton.)   Minnesota, (S. H. S.)

TRAGOCEPHALA, HARRIS.

*1. T. INFUSCATA, Harr., Report, 3d ed.; 181.   (1862.)
Gomphocerus infuscata, Uhl. in Harr., Report, 3d ed.; 181.   (1862.)

Mass., (H. Coll., Sanborn, Shurtleff.)   Maine, (Packard.) N. Hampshire, (H. Coll.)   Conn., (Norton.)

*2. T. VIRIDIFASCIATA, Harr., Report, 3d ed.; 182. Pl. 3, fig. 2.   (1862.)

For synonymy see Harris's Report; to which add:—

T. radiata, Harr., Report, 3d ed.; 183.   (1862.)
Locusta viridifasciata, Harr., Cat. Ins. Mass.; 56. (1835.)
Mass., (H. Coll., Shurtleff, Sanborn.)   Maine, (Packard.)
Conn., (Norton.)   Maryland, (Uhler.)

ARCYPTERA, SERVILLE.

Stetheophyma, Fischer Fr.
* 1. A. LINEATA, nov. sp.

Vertex of the head broad, slightly swollen at front border of the eye, apex docked, edge raised to a ridge, with a medial ridge extending over the whole top of the head; foveolæ small, shallow, triangular; lateral carinæ parallel in anterior half of pronotum, somewhat divergent behind, not so high as the medial, and much broken; wing-covers long and slender, with no swollen curves, the costal border not so prominent near the base as is usual in this genus; pronotum rugose.

Dark brown; a narrow curved dark line extends from the upper border of the eye to the lateral carinæ of the pronotum, and is the inner limit of a broad brownish yellow band which extends from the eye to the lateral carinæ, whence it continues backwards along the carinæ; below this upon the upper border of the side extends another broad black band from the eye to hind edge of pronotum; the medial carina is black; the wing-covers have the costal edge dark, beneath which is a yellow streak extending from base to the costal border at about two thirds the distance to the apex; beneath this is a band, narrow and black at base, broadening till it occupies the whole width of wing-cover, and becoming brown toward the tip, while the inner border is yellowish brown; wings dusky, the internal half with a yellowish tinge; legs dark brown; hind femora black on the outer and inner surface, reddish brown above, coral red below, with a white spot near apex, and the tip black; hind tibiæ yellow with black spines, with the base and tip black, and a dusky annulation at the upper limit of the spines.

Length of body, ♂ 1 in., ♀ 1.4 in.; of wing-cover, ♀ 1.12 in.; breadth of wing-cover in middle, ♀ .22 in.; length of hind femora, .72 in. 1 ♂, 1 ♀.

Mass., (Sanborn.)

* 2. A. PLATYPTERA, nov. sp.

Vertex of head much as in *A. lineata,* with the apex more rounded, and the edge scarcely prominent; foveolæ only discernible as faint depressions, slightly longer than in *A. lineata;* sides of pronotum regularly but slightly divergent; medial carina more prominent; pronotum rugose as in *A. lineata;* wing-covers short and broad, costal border considerably swollen near the base, internal border full.

Dark reddish brown, marked on head and prothorax similarly to *A. lineata,* but with the colors much suppressed; wing-covers uniform pale brownish, transparent; wings transparent, colorless, with the nervures of the anterior half black.

Length of body, 1.2 in.; of wing-covers, .83 in.; breadth of wing-covers in middle, .21 in.; length of hind femora .68 in. 1 ♀.

New England, (Agassiz.)

* 3. A GRACILIS, nov. sp.

Vertex of the head as in *A. lineata,* except that it is more pointed, narrower, and more rounded at the tip; foveolæ long and narrow, triangular, rather deep; pronotum as in *A. lineata;* wing-covers much as in *A. platyptera,* though the costal border is not swollen so much.

Dark brown; markings of head and thorax much as in *A. lineata,* though not so distinct; the band on the upper border of the sides of pronotum behind the eye is narrow, instead of broad; wing-covers uniform dusky brown, except the internal border which is yellowish brown; wings as in *A. lineata;* hind femora reddish, black at tip; hind tibiæ as in *A. lineata*, with the markings rather more distinct.

Length of body, .85 in.; of wing-covers, .78 in.; breadth of wing-covers in middle, .22 in.; length of hind femora, .52 in. 5 ♂.

Maine, (Packard.)   Red River Settlements, British America, (S. H. S.)

PEZOTETTIX, BURMEISTER.

1. P. BOREALIS, nov. sp.

Vertex of the head with a broad longitudinal furrow in advance of the middle of the eyes; sides of pronotum very nearly parallel, slightly wider at hind border which is arcuate; medial carina slightly higher than lateral, not prominent; wing-covers longer than wings, not quite reaching the extremity of the abdomen.

Dark brown, darkest above; a broad black band behind the eye, extending over the upper portion of the sides of pronotum to the hind border; front dark yellowish brown; mouth-parts dirty yellowish; legs yellowish brown; hind femora streaked with black, with the tip black; hind tibiæ reddish, with a faint paler annulation near base, the spines tipped with black; wing-covers dirty yellowish brown, spotted irregularly with darker brown; wings colorless, a little dusky on costal border.

Length of body, .65 in.; of wing covers, .4 in.; of hind femora, .4 in.

Minnesota, (S. H. S.)   Saskatchewan River, British America, (S. H. S.)   Lake Winnipeg, (S. H. S.)   Anticosti, Gulf St. Lawrence, (Verrill.)

CALOPTENUS, SERVILLE (*emend.*)

*1. C. FEMUR-RUBRUM, Burm., Handb. d. Ent.; II. 638. (1838.)

C. femur-rubrum, Uhler in Harr., Report, 3d ed.; 174. (1862.)

Acrydium femur-rubrum, Harr., Cat. Ins. Mass.; 56.
(1835.)
"   "   Harr., Report, 3d ed.; 174.
fig. 80.   (1862.)
For further synonymy see Harris's Report.

Mass., (H. Coll., Agassiz, Shurtleff, Miss Edmands, Sanborn, S. H. S.) Maine, (Packard.) Connecticut, (Norton.) S. Illinois, (Thomas.) Minnesota, (S. H. S.) Red River Settlements, (S. H. S.) Nebraska, (Mus. Comp. Zoöl.).

* 2. C. PUNCTULATUS, Uhler Mss.

This species is very closely allied to *C. femur-rubrum*, from which it is to be distinguished by the greater prominence of the front; by the greater globosity of the eyes; by the markings of the wing-covers and hind legs, and the sculpture of the valves of the ovipositor; the wing-covers are of the same color as in *C. femur-rubrum*, with the square dark spots not limited to a medial band, but found equally above and below this, extending nearly to the tip; the hind femora have upon the outside alternate, transverse, straight bands of black and brownish-yellow, three of each in number; hind tibiæ brownish-red with black spines, with a narrow black annulation at the base, followed by a wider yellowish one; the upper valves of the ovipositor are not curved so deeply on their upper surface, nor so upturned and pointed at their tip as in *C. femur-rubrum*; the lower valves, too, are much straighter, bent downwards scarcely at all at their tip, and the lateral tooth, so apparent in *C. femur-rubrum*, is here almost obsolete.

Length of body, 1.1 in.; of wing-covers, .69 in.; of hind femora, .48 in.; 1 ♀. I have seen only a single specimen sent me by Mr. Uhler under the above name.

Maine, (Packard.).

* 3. C. BIVITTATUS, Uhler in Say, Ent. of N. Am. (Ed. LeConte); II. 238. (1859.)
C. bivittatus, Uhler in Harr., Report, 3d ed.; 174. (1862.)

Gryllus bivittatus, Say, Journ. Ac. Nat. Sc. Phil.; IV. 308. (1825.)
Locusta leucostoma, Kirby, Faun. Bor. Am. Ins.; 250. (1837.)
Caloptenus femoratus, Burm., Handb. d. Ent.; II. 638. (1838.)
" " Burm., *teste* Erichson, Archiv. f. Nat.; IX. 229. (1843.)
Acrydium sanguinipes, Harr., Cat. Ins. Mass.; 56. (1835.)
" *flavovittatum*, Harr., Report, 3d ed.; 173. (1862.)

Mass., (H. Coll., Agassiz, Shurtleff, Miss Edmands, Sanborn.) Maine, (Packard.) Conn., (Norton.) Maryland, (Uhler.) Texas, (Uhler.) Nebraska, (Mus. Comp. Zoöl.) S. Illinois, (Thomas.) Minnesota, (S. H. S.) Lake Winnipeg, (S. H. S.)

ACRIDIUM, GEOFFROY.

* 1. A. ALUTACEUM, Harr., Report, 3d ed.; 173. (1862.)
A. torvum, Say Mss. and Harr., Cat. Ins. Mass.; 56. (1835.)
A. rusticum, Burm., Handb. d. Ent.; II. 633. (1838.)
(Not *Gryllus rusticus*, Fabr., Ent. Syst.; 292.)

Martha's Vineyard, Mass., (H. Coll.) Conn., (Norton.) 3 specimens.

2. A. AMERICANUM.
Gryllus americanus, Drury, Ill.; II. App.* Descr. and fig., I. 128. Pl. 49, fig. 2. (1773.)
Locusta tartarica, Westwood in Drury, Ill.; I. 121. Pl. 49, fig. 2. (1837.)
(Not *Gryllus tartaricus*, Linn., &c.)
N. Carolina, (H. Coll.) Southern States, (Mus. Comp.

---

* According to Westwood in his edition of Drury; the only copies of the old edition of Drury, which I have seen, have no appendix in any volume.

Zoöl.) Florida, (Norton.) Alabama, (H. Coll.) Texas, (Mus. Comp. Zoöl.) S. Illinois, (Thomas.)

3. A. OBSCURUM, Burm., Handb. d. Ent.; II. 632. (1838.)
Gryllus obscurus, Fabr., Syst. Ent. Suppl.; 194. (1798.)
This species has much stouter legs than the others.
Texas, (Mus. Comp. Zoöl., Uhler.)
("The *obscurum*? F. of H. Cat. is not in cabinet of T. W. H." Harris Mss.)

* 4. A. RUBIGINOSUM, Harris Mss.

" Light rust-red; face with four elevated lines, the two lateral ones not so distinct as in *A. alutaceum*; thorax with a very distinct ridge along the middle; wing-covers opaque, rather paler on the overlapping portion than elsewhere, with a projection on the outer margin near the shoulder; wings transparent and glassy, slightly reddish towards the tip and netted with blackish veins; hindmost thighs reddish within and without, the whitish part bounded on both sides by a row of distant black dots, and crossed like a herring-bone with reddish lines; knees of the same legs with a curved black line on each side, spines of the shins white tipped with black. Length, 1¾ in.; expanse of wings, nearly 3 inches. So. Carolina, from Mr. Nuttall and Dr. Holbrook." Harris Mss.

Dr. Harris does not mention the faint dark spots on the wing-covers, similar in disposition to those on *A. alutaceum*; there is some variation in the elevation of the medial carina of the pronotum, some individuals showing it but indistinctly.

Cape Cod, (Sanborn, S. H. S.) Conn., (Norton.) Southern States, (Mus. Comp. Zoöl.) Alabama, (H. Coll.) So. Carolina, (H. Coll.)

ŒDIPODA, LATREILLE.

I have here included all the species which would be embraced in the old genus *Œdipoda* of Latreille; it is easy to see that it should be divided, and especially that Œ. Carolina, Burm., and Œ. sordida, Burm., should each be separated from it, but the desire of having fuller material before attempting such a task prevents my undertaking it at present.

*1. Œ. CAROLINA, Burm., Handb. d. Ent.; II. 643.
(1838.)
Œ. carolina, Serv., Hist. Nat. d. Orth.; 722. (1839.)
"   "   Uhl. in Harr., Report, 3d ed.; 176. (1862.)
Locusta caroliniana etc. Catesby, Nat. Hist. of N. Car.; II. 89. Tab. 89. (1743.)
Gryllus (Locusta) carolinus, Linn., Syst. Nat.; II. 701.
(1767.)
"   "   "   Stoll, Spectres, &c. Pl. XVIII. b. fig. 68. (1813.)
Gryllus carolinus, Fabr., Syst. Ent.; 291. (1775.)
"   "   "   Spec. Ins.; I. 368. (1781.)
"   "   "   Mant. Ins.; I. 238. (1787.)
"   "   "   Ent. Syst.; II. 58. (1743.)
Acridium carolinum, De Geer, Mem.; III. 491. Pl. 41, figs. 2, 3. (1773.)
"   "   Oliv., Enc. Meth.; IV. 225. (1791.)
Acridium carolinianum, Pal. de Beauv., Ins.; 147. Pl. IV, fig. 6. (1805.)
Locusta carolina, Harr., Cat. Ins. Mass.; 56. (1835.)
"   "   Harr., Report, 3d ed.; 176. Pl. 3, fig. 3. (1862.)

Mass., (H. Coll., Miss Edmands, Shurtleff, Agassiz, Sanborn, S. H. S.) Maine, (Packard.) Conn., (H. Coll., Norton.)

*2. Œ. PHOENICOPTERA, Germ. in Burm., Handb. d. Ent.; II. 643. (1838.)
"   "   Germ., *teste* Erichson, Archiv. f. Nat. IX. 229. (1843.)
Œdipoda obliterata, Germ. in Burm., Handb. d. Ent.; II. 643. (1838.)

Locusta apiculata, Say Mss., and Harr., Cat. Ins. Mass.;
56. (1835.)
Locusta corallina, Harr., Report, 3d ed.; 176. (1862.)
Mass., (H. Coll., Shurtleff, Sanborn, S. H. S.) Maine, (Packard.) Conn., (Norton.)

3. Œ. DISCOIDEA, Serv., Hist. Nat. d. Orth.; 724. (1835.)
Acridium tuberculatum, Pal. de Beauv., Ins.; 145. Pl. IV, fig. 1. (1805.)
(Not *Gryllus tuberculatus*, Fabr.)
N. Carolina, (H. Coll.) Southern States, (Mus. Comp. Zoöl.)

* 4. Œ. RUGOSA, nov. sp.

This species is closely allied to *Œ. discoidea*. The head and thorax are dark brown; two yellowish bands run from behind the eye backwards and inwards, nearly or quite meeting one another a little in advance of the middle of the pronotum, where they diverge and strike the hinder edge of the pronotum at the outer angles; there are two yellowish spots, one below the other, on the sides of the pronotum; the wing-covers are marked much as in *Œ. discoidea*, but the dark blotches are larger and fully as abundant, generally occupying the larger portion of the wing, so that it might better be described as very dark brown with light blotches; the tip of the wing-cover is pellucid, nearly free of spots; the wings are as in *Œ. discoidea*, with the basal color pale-yellowish instead of red, and the apical portion less dusky than there.

Length of body, ♂ 1.1 in., ♀ 1.4 in.; expanse of wings, ♂ 1.9 in., ♀ 3 in.; depth of wings, ♂ .5 in., ♀ .7 in. 4 specimens.

Mass., (Agassiz.) Maine, (Packard.)

* 5. Œ. XANTHOPTERA, Germ. in Burm., Handb. d. Ent.; II. 643. (1838.)

Mass., (H. Coll., Agassiz, Sanborn, Shurtleff,) Missouri, (S. H. S.)

*6. Œ. SULPHUREA, Burm., Handb. d. Ent.; II. 643. (1838.)
" " Uhl. in Harr., Report, 3d ed.; 177. (1862.)
Gryllus sulphureus, Fabr., Spec. Ins.; I. 369. (1781.)
" " " Mant. Ins.; I. 239. (1787.)
" " " Sys. Ent.; II. 59. (1793.)
Acridium sulphureum, Oliv., Enc. Meth.; IV. 227. (1791.)
" " Pal. de Beauv., Ins.; 145. Pl. 4, fig. 2. (1805.)
Locusta sulphurea, Harr., Cat. Ins. Mass.; 56. (1835.)
" " " Report, 3d ed.; 177. Pl. 1, fig. 6. (1862.)

Mass., (H. Coll., Sanborn, Shurtleff.) Maine, (Packard.) Conn., (Norton.)

Œ. sulphurea differs from *Œ. xanthoptera* in its smaller size (the males of *Œ. xanthoptera* equalling in size the females of *Œ. sulphurea*), in the squareness and greater size of the foveolæ of vertex, in the direction of the edges of the ridge down the front (which in *Œ. sulphurea* are brought together at the vertex), in the direction of the hind-border of the pronotum (which is much more angulated in *Œ. xanthoptera*), in the greater depth of the wing in *Œ. xanthoptera*, and also in the band of the wing, which in both species has the inner border turned considerably inwards close to the costal border, and then outwards again just before the edge, but which in *Œ. sulphurea* extends inwards fully half way to the base of the wing, while in *Œ. xanthoptera* it does not reach one quarter the distance.

*7. Œ. ÆQUALIS, Uhl. in Harr., Report, 3d ed.; 178. (1862.)
Gryllus æqualis, Say, Journ. Acad. Nat. Sc. Phil.; IV. 307. (1825.)
" " Say, Ent. of N. Am. (ed. Le Conte); II. 237. (1859.)

Locusta æqualis, Harr., Cat. Ins. Mass.; 56. (1835.)
"   "   "   Report, 3d ed.; 178.   (1862.)
Mass., (H. Coll., Agassiz, Shurtleff, Sanborn, S. H. S.) Conn., (Norton, S. H. S.) Minnesota, (S. H. S.) Red River, British Am. (S. H. S.)

* 8. ŒE. VERRUCULATA.

Locusta verruculata, Kirby, Faun. Bor. Am. Ins.; 250.
(1837.)
"   *latipennis*, Harr., Report, 3d ed.; 179.   (1862.)
Œdipoda latipennis=æqualis, Uhl. in Harr., Report, 3d
ed.; 178.   (1862.)
Mass., (H. Coll., Sanborn, Shurtleff, Agassiz, S. H. S.) N. Hampshire, (H. Coll.) White Mts., (Agassiz.) Maine, (Packard.) Lake Winnipeg, (S. H. S.) Saguenay River, Canada East, (Norton.)

Œ. *verruculata* differs from Œ. *æqualis* in the following particulars: in Œ. *æqualis* the black band across the middle of the wings is broad, its outer edge as well as the inner distinct, the outer border at first straight, then well rounded, curving inwards where it approaches the outer border; beyond the band the wing is pellucid, with black veins, not cloudy, and at the tip there is either a dusky patch or irregularly clustered square blackish spots. In Œ. *verruculata* the inner border of the band is more wavy and is illy defined; the outer border is straight, and where it approaches the outer border of the wing, is turned slightly outwards instead of inwards, and is frequently very indistinct, being merged into the more or less dusky space beyond it, which increases in cloudiness to the tip, where it is as dark as the band. The band itself is quite narrow in the middle, so that it might be said to be made up of two triangular patches which meet and merge in the middle. The broadest band I have seen in Œ. *verruculata*, is not more than half the width of the narrowest in Œ. *æqualis*. In Œ. *æqualis* the hind tibiæ are either wholly coral-red or have a pale yellowish annulation at the base. In Œ. *verruculata* the tibiæ have the base and apex black, with the middle half yellowish or plumbeous, with generally a dusky annulation in the middle.

* 9. Œ. MARITIMA, Uhl. in Harr., Report, 3d ed.; 178. (1862.)
Locusta maritima, Harr., Report, 3d ed.; 178.   (1862.)
Sea-shore of Mass., (H. Coll., Sanborn, Miss Edmands, S. H. S.) Conn., (Norton.)

* 10. Œ. MARMORATA, Uhl. in Harr., Report, 3d ed.; 179. (1862.)
Locusta cerineipennis, Harr., Cat. Ins. Mass.; 56. (1835.)
"   *marmorata*, Harr., Report, 3d ed.; 179.   (1862.)
Mass., (H. Coll., A. Agassiz, Shurtleff, Sanborn.)

* 11. Œ. EUCERATA, Uhl. in Harr., Report, 3d ed.; 180. (1862.)
Locusta eucerata, Harr., Cat. Ins. Mass.; 56.   (1835.)
"   "   "   Report, 3d ed.; 180.   (1862.)
Mass., (H. Coll., Shurtleff, Sanborn, S. H. S.)   Conn., (Norton, S. H. S.)

* 12. Œ. PELLUCIDA, nov. sp.

Ash brown; face reddish brown; antennæ yellowish at base, dark brown toward tip; a triangular black spot behind the eye, the apex touching it; a quadrate transverse black spot on the anterior upper portion of the sides of pronotum; pronotum above, sometimes with a dark band down the middle; wing-covers with the basal half dark brown, with small yellowish spots and transverse streaks, especially on front border; apical half clear, with dark brown rounded spots, prevalent along the middle, decreasing in size toward the tip; when closed, the upper surface is dark brown, with a rather broad yellowish vitta along each angle on the upper surface; wings pellucid, with black nervules; legs dark brown, the hind femora yellowish or reddish-brown, with two or three rather broad diagonal dark-brown streaks, dark at the apex; hind tibiæ yellowish-brown, reddish toward the tip, with a very narrow, generally faint, annulation of dark-brown at the base; spines tipped with black.

Length of body, ♂ .65 in., ♀ 1 in.; spread of wings, ♂ 1.3 in., ♀ 1.6 in.; depth of wings, ♂ .33 in., ♀ .4 in.

Mass., (Miss Edmands, Agassiz, Sanborn, S. H. S.) Vermont, (S. H. S.) Maine, (Packard.) Conn., (Norton.)

* 13. Œ. SORDIDA, Burm., Handb. d. Ent.; II. 643. (1838.)

Locusta periscelidis, Say Mss., and Harr., Cat. Ins. Mass.; 56. (1835.)

Locusta nebulosa, Harr., Report, 3d ed.; 181. (1862.)

Œdipoda nebulosa, Uhl. in Harr., Report, 3d ed.; 181. (1862.)

Mass., (H. Coll., Agassiz, Shurtleff, Sanborn, S. H. S.) Maine, (Packard.) Conn., (Norton.)

14. Œ. COSTALIS, nov. sp.

Brownish-yellow, profusely mottled with reddish-brown; a broad yellowish band extends from each eye to the opposite outer posterior angle of the pronotum, crossing at the middle of the pronotum, bordered outside with a broad band of black, narrowing posteriorly to a line, and inside, behind the intersection, by a broad black band, which has another narrow short black line parallel to it at the hind border of the pronotum; medial carina of pronotum equal, rather sharp, not high; wing-covers much as in *Œ. sordida;* wings pellucid, with a faint cloudy patch at the middle of the outer border, and a dark streak along the costal border toward the apex; hind femora yellowish, with two transverse dark annulations, the tips dark brown; hind tibiæ bluish with black spines and a yellowish annulation at base. This species differs from *Œ. sordida*, in the markings and lowness of the ridge of pronotum, and in the shorter and fuller wings.

Length of body, 1.05 in.; spread of wings, 1.75 in.; depth, .46 in. 1 ♀.
Texas, (Mus. Comp. Zoöl.)

TETTIX, LATREILLE (emend.)

See *Tettigidea* and *Batrachidea*.

* 1. T. GRANULATA.

Acrydium granulatum, Kirby, Faun. Bor. Am. Ins.; 251. (1837.)
Tetrix ornata, Harr., Cat. Ins. Mass.; 57. (1835.)
"   "   Harr., Report, 3d ed.; 186. (1862.)
(Not *Acrydium ornatum*, Say.)

This species may be distinguished from *T. ornata*, by its longer pronotum and greater size, and also by the prominence of the vertex, which advances considerably in front of the eyes, having the front border angulated; in this latter feature it may also be similarly distinguished from *T. cucullata* and *T. rugosa*, as also by its narrower pronotum; the males are much narrower than the females.

Mass., (H. Coll., Sanborn, Shurtleff, S. H. S.) Maine, (Packard.) N. Hampshire, (H. Coll.) Minnesota, (S. H. S.)

* 2. T. ORNATA.

Acrydium ornatum, Say, Amer. Entom.; I. Pl. V. (1824.)
"   "   Say, Ent. of N. Am. (ed. LeConte);
                I. 10. Pl. V. fig. 1. (1859.)
Tetrix arenosa, Burm., Handb. d. Ent.; II. 659. (1838.)
"   *dorsalis*, Harr., Report, 3d ed.; 186. (1862.)
"   *quadrimaculata*, Harr., Report, 3d ed.; 186. (1862.)
"   *bilineata*, Harr., Report, 3d ed.; 186. (1862.)
"   *sordida*, Harr., Cat. Ins. Mass.; 57. (1835.)
"   "   Harr., Report, 3d ed.; 187. (1862.)
(Not *Tetrix ornata*, Harr., Cat. and Report.)

This species is smaller than *T. granulatum*, has the vertex but little thrust forward in advance of the eyes, and the front border nearly straight instead of angulated; the pronotum is shorter than in the preceding, and the wings are smaller; both this and the preceding species have almost every conceivable variation of ornamentation, upon which almost exclusively Harris established his specific differences, but as Uhler has remarked, "color and style of marking is of very little value in separating the species of Tettix."

Mass., (H. Coll., Sanborn, Shurtleff.) N. Hampshire, (H. Coll.) Maine, (H. Coll., Packard.) Vermont, (S. H. S.) Conn., (Norton.) S. Illinois, (Thomas.) St. Louis, Missouri, (Sanborn, S. H. S.)

* 3. T. TRIANGULARIS, nov. sp.

Allied to *T. ornata*, and agreeing with it in ornamentation, in the character of the vertex, the prominence of the eyes, but differing in the length of the pronotum and wings; as in both of the preceding species, the pronotum and wings are of equal length, but in this the pronotum is scarcely longer than the body, and is not produced backward into such a slender point, the sides being straighter; the breadth is contained three times in the length; it is a smaller species than the preceding.

Length of pronotum, .17 in. 2 ♂, 2 ♀.

Mass., (S. H. S.) Maine, (Packard.) N. Hampshire, (H. Coll.)

* 4. T. CUCULLATA, Burm., Handb. d. Ent.; II. 658. (1838.)

Differs from *T. granulata*, which it most resembles, in having the vertex very narrow, slightly less than the diameter of the much inflated eyes, the front cut off square, and slightly hollowed, not projecting outward so far as the eyes; the pronotum is broader and more compact over the thorax, more suddenly sloped off behind and extending backwards nearly twice the length of the abdomen, the wings overreaching slightly; the punctures upon the wing-covers are of the same size, but not so deep as in *T. granulata*.

Length of pronotum, ♂, .4 in., ♀ .5 in. 2 ♂, 2 ♀.

Mass., (S. H. S.) Missouri, (Mus. Comp. Zoöl., Sanborn.)

5. T. RUGOSA, nov. sp.

Closely allied to *T. cucullata*, agreeing with it in general form and size, shape and length of the pronotum, and length of the wings. The front border of the vertex is as in *T. cucullata*, but it is broader, and the eyes are scarcely as prominent. The whole surface of the pronotum, instead of being delicately granulated as in *T. cucullata*, with the medial and marginal carinæ faint, has these carinæ quite prominent, and the whole surface rugose, deeply scarred and pitted, with irregular granulated depressions; the wing-covers are punctured as in *T. granulata*.

Length of pronotum, .54 in. 1 ♀.

N. Florida, (Norton.)

TETTIGIDEA, Nov. gen. (τέττιξ, ιδέα.)

This genus when compared with *Tettix* will be found to differ in having a more robust and clumsy form, a larger head, more swollen upon the top, and less sloping down the front, the medial ridge in front more prominent, the antennæ consisting of twenty-two joints, which are cylindrical and not flattened; in the joints of the maxillary palpi, which here have the fourth joint much larger at the apex than at the base, somewhat swollen, with a sharp medial external ridge, and the fifth much swollen, flattened, with a faint similar ridge, and slightly docked at the tip; as in *Tettix* and *Batrachidea* the first joint is longer than broad, cylindrical, the second slightly shorter than broad, cylindrical, both together equalling the third, which is of the same length as the fourth or fifth, and cylindrical; the lower anterior angle of the sides of pronotum, which is angulated and bent inwards in *Tettix*, is here rounded and straighter; the lateral carinæ are not so prominent as there, or so strongly bent inwards in advance of the broader portion; the front border is thrust forward at an angle partially concealing the head; the prosternum is very strongly folded transversely, forming a very deep, sharp, angulated groove, which in *Tettix* is not nearly so deep, nor are its sides so nearly approximated; wing-covers considerably longer and narrower than in *Tettix*. This genus further differs from *Tettix* in that there is a small circular swollen space devoid of facets, set off from the upper inner border of the eye. The same is true of *Batrachidea*, but much more indistinctly, since it cannot be discovered without the aid of such a lens as will readily separate the facets of the eye.

*1. T. LATERALIS.

Acrydium laterale, Say, Am. Ent.; I. Pl. 5.   (1824.)
"   "   Say, Ent. of N. Am. (ed. Le Conte);
                     I. 10.   Pl. 5, figs. 2, 3.   (1859.)
Tetrix lateralis, Harr., Cat. Ins. Mass.; 57.   (1835.)
"   "   Harr., Report, 3d ed.; 187.   (1862.)
"   *polymorpha*, var. A, Burm., Handb. d. Ent.; II.
                     659.   (1838.)

Mass., (H. Coll., Sanborn.)   Maine, (Packard.)   N. Hampshire, (H. Coll.)   Conn., (Norton.)   S. Illinois, (Thomas.)

*1. T. POLYMORPHA.

Tetrix polymorpha, var. B, Burm., Handb. d. Ent.; II.
                     659.   (1838.)
Tetrix parvipennis, Harr., Cat. Ins. Mass.; 57. (1835.)
"   "   Harr., Report, 3d ed.; 187, fig. 82.
                                                  (1862.)

In this species the wings are almost abortive, and the pronotum extends only to the tip of the abdomen, while in *T. lateralis* the wings extend beyond the pronotum, which is itself much longer than the body; *T. polymorpha* is the more abundant species.

Mass., (H. Coll., Sanborn.) Maine, (H. Coll., Packard.) N. Hampshire, (H. Coll.) Conn., (Norton.) S. Illinois, (Thomas.) St. Louis, Missouri, (Sanborn.) Alabama, (H. Coll.)

BATRACHIDEA, SERVILLE.

This genus differs from *Tettix* in its more solid and compact form; in the larger head, the more distant eyes, the front less sloping; in the smaller number of joints in the antennæ, which have but twelve joints, while in *Tettix* there are thirteen or fourteen, generally the latter; in the shape of their joints, which are cylindrical instead of being flattened, and more swollen than in *Tettix;* in the maxillary palpi, which in *Tettix* has the fourth joint cylindrical, very slightly largest at the apex, and the fifth cylindrical and slightly swollen, while here the fourth is somewhat larger at the apex than at the base, broadly but faintly ridged outside, and the fifth swollen considerably, especially on anterior border, with a broad faint ridge outside; in the more swollen and crested summit of the head; in having the lower posterior lobes of sides of pronotum thrust downwards and outwards and but slightly backwards, and the lower anterior angle rounded; in having a very high arched median carina on pronotum, and the lateral carinæ only indicated in front; in having the front border of pronotum thrust forward over the head a little; in having on prosternum only a broad shallow rounded transverse hollowing; in having the notches on the under side of the first joint of posterior tarsi only very slight, instead of being prominent as in *Tettix* and in *Tettigidea;* in the shorter valves of the ovipositor; and in having stouter legs than in the two genera just mentioned. See also *Tettigidea*.

* 1. B. CRISTATA.

Tetrix cristata, Harris Mss.

Vertex projecting beyond the eyes, front border well rounded, a little angulated, the medial carina sharp, prominent, sloping downwards posteriorly, the front deeply notched immediately in front of the eyes; eyes rather prominent, scarcely more than half as broad as the vertex; the pronotum with sides neither swollen nor hollowed, of the length of the body; the medial carina high, regularly arched; the lateral border with two shallow grooves, one anterior, the other posterior, overlapping one another in the middle; the whole pronotum is minutely scabrous, and there is generally a dark quadrate or triangular spot on either side, above the terminal half of the wing-covers; wings reaching tip of pronotum.

Length of pronotum, .33 in.

Mass., (H. Coll., Shurtleff, Sanborn, S. H. S.) Maine, (H. Coll., Packard.) N. Hampshire, (H. Coll.) Conn., (S. H. S.)

* 2. B. CARINATA, nov. sp.

The head much as in *B. cristata*, with the eyes slightly larger and more prominent; the medial carina of the pronotum sharp, regularly arched, the pronotum extending backward a good ways behind the tip of the abdomen, a little upturned towards the tip, with slightly longer wings; the lateral grooves are narrower and less distinct than in *B. cristata*, and the upper surface is more coarsely scabrous than there; marked as in *B. cristata*.

Length of body, .32 in.; of pronotum, .43 in. 1 ♂, 2 ♀.
Mass., (Sanborn, S. H. S.)

EXPLANATION OF WOOD-CUTS.

Fig. 1, p. 424. Fore-tibia and tarsi of *Tridactylus apicalis*, Say ♂ (magnified.)
Fig. 2, p. 442. Hind-tarsi of *Udeopsylla nigra*, Scudd. (magnified 2 diameters.)
Fig. 3, p. 443. *a*. Hind-tarsi of *Daihinia brevipes*, Hald. (magnified 2 diameters.)
    b. Middle tarsi of *Daihinia brevipes*, Hald. (magnified 2 diameters.)
Fig. 4, p. 446. Wing-cover of *Phylloptera rotundifolia*, Scudd.
Fig. 5, p. 446. Wing-cover of *Microcentrum affiliatum*, Scudd.

Index to the Genera and Species in this Article.

Acridium, 466	Gryllotalpa, 426	Phaneroptera curvicauda, 448
Acridium alutaceum, 466	Gryllotalpa borealis, 426	Phylloptera, 444
americanum, 466	longipennis, 426	Phylloptera caudata, 445
obscurum, 467	Gryllus, 427	oblongifolia, 444
rubiginosum, 467	Gryllus abbreviatus, 427	rotundifolia, 445
Arcyptera, 462	angustus, 427	Platamodes, 417
Arcyptera gracilis, 463	bipunctatus, 432	Platamodes pennsylvanica, 417
lineata, 462	luctuosus, 427	unicolor, 417
platyptera, 463	neglectus, 428	Pycnoscelus, 421
Batrachidea, 478	niger, 428	Pycnoscelus obscurus, 422
Batrachidea carinata, 479	pennsylvanicus, 429	Spongophora, 415
cristata, 478	Hadenœcus, 439	Spongophora bipunctata, 415
Caloptenus, 464	Hadenœcus subterraneus, 440	Stenobothrus, 456
Caloptenus bivittatus, 465	Labia, 415	Stenobothrus æqualis, 459
femur-rubrum, 464	Labia minuta, 415	bilineatus, 460
punctulatus, 465	Microcentrum, 446	curtipennis, 456
Ceuthophilus, 433	Microcentrum affilatum, 447	longipennis, 457
Ceuthophilus Agassizii, 439	retinervis, 446	maculipennis, 458
brevipes, 434	thoracicum, 447	melanopleurus, 456
californianus, 438	Nemobius, 429	propinquans, 461
divergens, 439	Nemobius exiguus, 429	speciosus, 458
gracilipes, 439	fasciatus, 430	Stylopyga, 416
lapidicolus, 435	vittatus, 430	Stylopyga orientalis, 416
latens, 437	Œcanthus, 431	Tettigidea, 476
maculatus, 434	Œcanthus niveus, 431	Tettigidea lateralis, 477
niger, 437	punctulatus, 432	polymorpha, 477
scabripes, 436	Œdipoda, 467	Tettix, 474
stygius, 438	Œdipoda æqualis, 470	Tettix cucullata, 475
Uhleri, 435	carolina, 468	granulata, 474
Chloëaltis, 455	costalis, 473	ornata, 474
Chloëaltis conspersa, 455	discoidea, 469	rugosa, 476
punctulata, 455	eucerata, 472	triangularis, 475
viridis, 455	maritima, 472	Thyreonotus, 453
Conocephalus, 449	marmorata, 472	Thyreonotus dorsalis, 454
Conocephalus crepitans, 450	pellucida, 472	pachymerus, 453
ensiger, 449	phœnicoptera, 468	Tragocephala, 461
obtusus, 450	rugosa, 469	Tragocephala infuscata, 461
robustus, 449	sordida, 473	viridifasciata, 461
uncinatus, 450	sulphurea, 470	Tridactylus, 424
Cryptocercus, 419	verruculata, 471	Tridactylus apicalis, 425
Cryptocercus punctulatus, 420	xanthoptera, 469	minutus, 425
Cyrtophyllus, 444	Opomala, 454	terminalis, 425
Cyrtophyllus concavus, 444	Opomala brachyptera, 454	Tropidischia, 440
perspicillatus, 444	Orchelimum, 452	Tropidischia xanthostoma, 441
Daihinia, 443	Orchelimum agile, 453	Udeopsylla, 442
Daihinia brevipes, 443	concinnum, 452	Udeopsylla nigra, 443
Diapheromera, 423	glaberrimum, 453	robusta, 442
Diapheromera femorata, 423	longipennis, 453	Xiphidium, 451
Ectobia, 418	vulgare, 452	Xiphidium brevipennis, 451
Ectobia flavocincta, 419	Periplaneta, 416	ensifer, 451
germanica, 418	Periplaneta americana, 416	fasciatum, 451
lithophila, 418	Pezotettix, 464	
	Pezotettix borealis, 464	
	Phaneroptera, 448	

www.ingramcontent.com/pod-product-compliance
Lightning Source LLC
Chambersburg PA
CBHW022147090426
42742CB00010B/1420